Artificial Organ Engineering

Maria Cristina Annesini · Luigi Marrelli
Vincenzo Piemonte · Luca Turchetti

Artificial Organ Engineering

Maria Cristina Annesini
Department of Chemical Engineering
  Materials and Environment
University "La Sapienza" of Rome
Rome
Italy

Luigi Marrelli
Faculty of Engineering
University "Campus Bio-medico" of Rome
Rome
Italy

Vincenzo Piemonte
Faculty of Engineering
University "Campus Bio-medico" of Rome
Rome
Italy

Luca Turchetti
ENEA- Italian National Agency for New
  Technologies, Energy and Sustainable
  Economic Development
Rome
Italy

ISBN 978-1-4471-7383-0   ISBN 978-1-4471-6443-2   (eBook)
DOI 10.1007/978-1-4471-6443-2

The information presented in the book is addressed to engineers and is not intended to be directly used to take any medical decision.

Printed on acid-free paper

This Springer imprint is published by Springer Nature
The registered company is Springer-Verlag London Ltd.

Introduction

The history of medicine has always been characterized by the attempt to treat a wide range of diseases, some very serious and with high mortality, and others debilitating and detrimental for the quality of life of patients. With the increase of life expectancy, organ failure has become quite common, making the problem of degeneration of some body parts (organs, joints etc) increasingly critical. Therefore, the possibility of replacing these parts, represents an interesting opportunity for increasing life duration and improving its quality.

Substitution of a part of the human body can be achieved by transplantation from a human or animal donor. Tissues for transplantation can be obtained from the recipient's own body (autotransplantation). Tissues or organs can be taken from a different living or dead compatible human donor (allotransplantation) or from an animal (xenotransplantation). Unfortunately, while the population of patients requiring organs continues to increase, the lack of an adequate number of donors, along with biological and ethical problems connected with allotransplantation and xenotransplantation, makes organ transplantation still inadequate and the number of patients on the waiting lists is growing rapidly. A possible alternative to transplantation consists in the use of artificial and bio-artificial organs. The availability of devices able to substitute, or at least support, damaged vital functions can allow the patient to be kept alive a long time or, at least, until either a transplantation is possible or the physiological activity of the native organ is restored. Furthermore, artificial organs play a key role in enhancing a patient's quality of life. However, the current state of development in the fields of biotechnology and bioengineering does not allow all organs and tissues to be available.

At present, several extracorporeal artificial assist devices are available and in use, such as the artificial kidney, whereas only few implantable devices are approved for clinical use. In the last decades, biomedical engineering has made great strides in this field with the support of nanotechnology, microelectronics, and biology and with the significant contribution of the fundamentals of chemical engineering such as thermodynamics, kinetics, and transport phenomena; therefore, we can imagine that in the future, the number of miniature artificial organs for permanent implantation will increase. A comprehensive definition considers artificial organs "any equipment, device, or material, directly or indirectly interfaced with living tissue and used to substitute, partly or entirely, or to strengthen functions of a natural organ or of any other part of the body badly working or lacking." This definition, drawn from a conference of the National Institute of Health, considers as artificial organs both devices performing physical–chemical functions (such as artificial kidney, blood oxygenator, and artificial liver.) and electromechanical devices (such as pacemakers, heart valves, artificial hands, and orthopedic prostheses) or aesthetic parts (such as mammary prostheses). A different approach distinguishes between artificial (and bio-artificial) organs and prostheses, defining the former as devices substituting or supporting any physical–chemical function of the body and the latter as devices designed only for mechanical or electromechanical functions.

The earliest artificial organs were mostly based on mechanical technologies. The first artificial kidney, which marks the beginning of artificial organ history, was basically a blood filter aimed at removing waste material from the body. Likewise, artificial hearts and ventricular assist devices (VADs) were all based on pump and valve technology. However, biomedical researchers have quickly realized that most human organs cannot be substituted by artificial ones mimicking only their mechanical functions. Endocrine organs, for example, are exceedingly complex in their functions to be artificially reproduced at the present state of scientific and technological knowledge. A typical case is represented by the liver, which carries out many biological functions, among which blood detoxification and synthesis of biomolecules essential in metabolism are only the most well-known.

While blood detoxification can be fairly performed by the use of membranes, synthesis mechanisms and other hepatic functions are still far from being reproduced by an artificial liver. Today, many artificial devices are rough simplifications of the biological original and are able to reproduce only some of the vital functions. For these reasons, biomedical research has concentrated its efforts on the development of hybrid systems, coupling biological and artificial components. These devices, named bio-artificial organs, usually contain a bioreactor where cells or a tissue of the organ to be substituted carry out the functions of the native organ. Besides artificial and bio-artificial organs, a third approach, named neo-organs, is now emerging. This approach is closely connected with tissue engineering and is based on growing, over suitable biodegradable supports (scaffold), cells of the tissue to be produced or stem cells. Tissues with various three-dimensional structures can be currently produced with this technique. Very good results have been obtained in the production of bone and skin tissue, to be used in case of burns. Research is in progress in the field of nerves, muscles, and blood vessels.

As for market scenarios, according to a recent report published by the Transparency Market Research[1], the world market of artificial organs, including

---

[1]http://www.transparencymarketresearch.com

Artificial vital organs and medical bionics market (artificial heart, kidney, liver, pancreas and lungs, ear bionics, vision bionics, exoskeletons, bionic limbs, brain bionics and cardiac bionics)— Global Industry analysis, size, share, growth, trends and forecast, 2012–2018.

prostheses, is expected to grow at a compound annual growth rate (CAGR) of 9.2 %. Since in 2011 the artificial vital organs and medical bionics market were evaluated at about US$ 17.5 billion, the above CAGR value gives a forecast of US$ 32.3 billion in 2018. It is a substantial and continuously developing market that has a preeminent importance in technologically advanced countries, especially in the field of new devices. The global market of artificial organs is led by the artificial kidney, which made up 48 % of the global market in 2010. Its use is highly recommended as a short-/medium-term treatment, especially when used as a support while waiting for a kidney transplant. Industrial research is focused on the development of better membranes and on more efficient and cheaper production technology. Recently, promising steps forward have been taken in the field of artificial and bio-artificial livers. Devices such as MARS (Molecular Adsorbent Recirculating System) and ELAD are quite largely used to provide for the detoxifying needs of the organism. A bio-artificial liver could also provide the metabolic functions of natural liver. In coming years, technological advancements are expected in the field of an implantable bio-artificial pancreas with remarkable lucrative prospects connected with the current spreading of diabetes mellitus. Besides the improvement in safety and efficacy, an aspect to be taken into account is the reduction of production and operation costs of present and future devices, in order to make them affordable for a greater number of people, especially in developing countries. For most of these people, the dialysis treatment, diffusely used in advanced countries, is still a dream barred by poverty conditions.

A very important issue to be taken into account in the development and production of artificial organs concerns materials (biomaterials) to be used in contact with the tissues of the human body. These materials play a fundamental role in making a medical device safe for human use. The main requirement for a biomaterial is biocompatibility, i.e., the property of not causing toxic or damaging effects on biological systems and not activating the immune system. However, it is important to note that biocompatibility is not a property of the material alone, as it depends on the position in which the material is used or implanted and on the time of exposure to the biologic matter. From this point of view, it is possible that a material can be considered as biocompatible if used in extracorporeal devices, but not for internal use. Furthermore, biocompatibility is affected by the production process of the material and its state of cleaning and sterilization. The assessment of biocompatibility is regulated by ISO 10993 standards, which describe tests to be performed, in vitro or on animals, depending on the category of contact with human body. These tests, which must be carried out in specialized laboratories, concern toxicity, carcinogenicity, hemocompatibility, etc., and are sometimes very expensive (up to 100,000 €). Besides biocompatibility, other more conventional properties of a material to be used in artificial organs are mechanical strength and durability. For example, a normal heart beats about 40 million times a year. Therefore, parts of an artificial heart used to pump blood must be made of materials able to work for a long time without being deformed or broken. Another example refers to materials used in orthopaedic prostheses, whose average life is about 10–15 years.

In substituting a natural joint by an artificial one, two mobile parts touching each other must slide with a low friction coefficient and negligible tear effects. Suitable biomaterials must not involve the release of small debris with harmful effects since, besides wear effects, they activate the reaction of the immune system, which is often coupled with the release of enzymes that destroy the adjoining tissue. An additional problem connected with the use of biomaterials is the possible formation of biofilms. These are aggregates of bacteria which stick irreversibly to surfaces making multilayer settlements incorporated in a porous matrix that shelters the microorganisms from the attack of antibiotics. This problem can appear in catheters, in contact lenses, and, with serious effects, in heart valves. The formation of biofilms depends on the physical and chemical properties of the support surface, especially on its roughness and porosity, and on material hydrophobicity and chemical composition. Several groups of scientists and bioengineers are investigating solutions to prevent the formation of biofilms through specific coatings and surface treatments. Today, with the increasing chemical and biological knowledge, the point of view on the features of biomaterials is changing. For example, in the past, the greatest chemical and biological inactivity was required for a biomaterial, while now several very reactive materials are proving to be more suitable for some biomedical applications. Some materials, for example, form chemical bonds with the surrounding tissue, increasing the stability of prostheses. Other materials degrade and can be adsorbed onto the tissue when they are no longer necessary.

To conclude, it is very important to highlight that a proper design and operation of artificial and bio-artificial organs require a deep knowledge of fundamentals of chemical thermodynamics, transport phenomena, and chemical kinetics, besides anatomy and physiology. Most of the methods used in blood detoxification are based on physicochemical operations aimed at removing, in a short time, clinically important amounts of some substances without appreciable risk for the patient. Such separation operations are usually performed by selective membranes permeable to the toxic substances to be removed and impermeable to the essential compounds. Understanding of mass transfer across these membranes is therefore the basis for the design of a hemodialyzer. In hemoperfusion, toxic substances are removed by adsorption on solid adsorbents and solid–liquid phase equilibrium is involved. Furthermore, a rheology analysis is usually required in order to avoid blood cell damage. Likewise, the use of animal cells in bio-artificial organs requires the knowledge of the fundamentals of bioreactors, often coupled with mass and heat transfer processes.

The book is divided in two parts: The first one provides a presentation of the physical fundamentals involved in the technology of artificial and bio-artificial organs; the second one is devoted to the monographic presentation of the most important organ support and replacement devices based on mass transfer operations. More specifically, in the first part, mass transport phenomena are firstly discussed, from both a local and macroscopic point of view; separation processes widely used in artificial organs, i.e., separation based on transport through selective membranes and adsorption, are then presented; finally, the fundamentals of bioreactor engineering are covered, focusing on the interaction between bioreaction kinetics and transport phenomena. The three chapters of the second part are devoted to devices for blood oxygenation, renal replacement therapy, and liver support. For each device, a survey of commonly used solutions and the most promising developments is presented. Mathematical models to assess the performance are reported as fundamental tools for the quantitative description of clinical devices; nevertheless, the models proposed are kept simple to keep the focus on the essential features of each process.

Contents

Part I
Fundamentals

Chapter 1
Diffusion

1.1 Introduction

This chapter presents the fundamentals of diffusion processes, which are of paramount importance for the study of most types of artificial organs. Diffusion is a wide topic thoroughly covered in several textbooks [1], and a comprehensive treatise on this subject is beyond the scope of this book. Therefore, only a brief overview of the main concepts related to diffusion processes are presented here and the analysis is kept as simple as possible.

Only diffusion in dilute solutions will be considered, since the study of these mixtures allows to seize the main concepts about the physics of diffusion without having to deal with very involved mathematics. Moreover, many relevant diffusion problems related to artificial organs actually involve dilute solutions.

Let us consider a solution of a component A in a solvent B, with a non-uniform molar fraction of A (see Fig. 1.1); in such conditions, a *diffusive* flux (flow rate per unit area) of the component A in the direction z is observed, caused by random molecular motion, which tends to uniform the system composition. The second law of thermodynamics states that the diffusive flux of A must be directed toward the decreasing molar fraction of A in order to have an increase of entropy in the system:

$$J_{A,z} = f\left(\frac{dx_A}{dz}\right) \qquad J_{A,z} > 0 \text{ if } \frac{dx_A}{dz} < 0 \qquad (1.1)$$

There are experimental evidences that, for all the gases and almost all the liquids, the diffusive flux is proportional to the first derivative of the molar fraction of A:

$$J_{A,z} = -c\mathcal{D}_{AB}\frac{dx_A}{dz} \qquad (1.2)$$

where c is the molar density and the proportional coefficient $\mathcal{D}_{AB}$ is the diffusivity of A–B mixture. Equation 1.2 is the one-dimensional form of the Fick's law. If the molar

© Springer-Verlag London 2017
M.C. Annesini et al., *Artificial Organ Engineering*,
DOI 10.1007/978-1-4471-6443-2_1

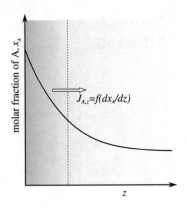

Fig. 1.1 Diffusion in a diluted mixture

fraction of A varies in all directions in space, an equation like (1.2) can be written for each coordinate and, if the fluid is isotropic, the proportionality coefficients are the same in all the directions. In this case, Fick's law may be written in a more synthetic vectorial form as[1]:

$$\mathbf{J}_A = -c\mathscr{D}_{AB}\nabla x_A \tag{1.3}$$

where $\mathbf{J}_A$ is the vector of the diffusive flux of A and ∇x_A is the gradient of the molar fraction of A. It is important to clearly state that while the second law of thermodynamics requires that the diffusive flux is directed toward the decreasing molar fraction of A, Fick's law—i.e., the proportionality between the diffusive flux and the molar fraction gradient—does not derive from any fundamental law, but it is only an empirical relation that has been proven to suitably describe many real systems.

For a dilute solution, the molar density can be considered constant in the system and the Fick's law may be rewritten in a more usual form as

$$J_{A,z} = -\mathscr{D}_{AB}\frac{dc_A}{dz} \quad \text{or} \quad \mathbf{J}_A = -\mathscr{D}_{AB}\nabla c_A \tag{1.4}$$

Furthermore, considering that the flux of A can be written as the product of the concentration times the velocity of the component A, the above equations may be rewritten in terms of *diffusion velocity* as:

$$u^*_{A,z} = \frac{J_{A,z}}{c_A} = -\frac{\mathscr{D}_{AB}}{x_A}\frac{dc_A}{dz} \quad \text{or} \quad \mathbf{u}^*_A = -\frac{\mathscr{D}_{AB}}{x_A}\nabla x_A \tag{1.5}$$

Diffusive flux (or diffusion velocity) refers to the motion of the component A with respect to the mixture, driven by the A concentration gradient; if all the mixture is

---

[1]∇s is the gradient of a scalar s. In rectangular coordinates, ∇s is a vector of components $\partial s/\partial x$, $\partial s/\partial y$ and $\partial s/\partial z$; for the expression of ∇s in other coordinates see the Appendix A, at the end of this chapter.

in motion with a velocity u_z, a drift flux superimposes to the diffusive flux and the total flux through a fixed surface is given by

$$N_{A,z} = J_{A,z} + c_A u_z \quad \text{or} \quad \mathbf{N}_A = \mathbf{J}_A + c_A \mathbf{u} \tag{1.6}$$

1.2 A Rigorous Approach

Let us consider pure convective flux in a binary mixture with uniform composition: All the components move at the same average velocity,[2] like a pure component:

$$\mathbf{u}_A = \mathbf{u}_B = \mathbf{u} \tag{1.7}$$

and the fluxes of all the components with respect to a fixed frame are given by:

$$\mathbf{N}_i = c_i \mathbf{u} \tag{1.8}$$

On the other hand, in a mixture with non-uniform mole fraction profile, the various chemical species move at different velocities $u_A \neq u_B$ (actually, this is what we refer to as *diffusion*) and the flux of each component in a fixed reference frame is given by

$$\mathbf{N}_A = c_A \mathbf{u}_A \tag{1.9}$$

The motion of A can also be considered with respect to the motion of the mixture, i.e., in a reference frame moving at the average velocity of the mixture: thus, the *diffusion velocity* is defined as

$$\mathbf{u}_A^* = \mathbf{u}_A - \mathbf{u} \tag{1.10}$$

and the *diffusive flux* is defined as the molar flux relative to the molar average velocity:

$$\mathbf{J}_A = c_A (\mathbf{u}_A - \mathbf{u}) \tag{1.11}$$

Figure 1.2 presents an overview of convective and diffusive transport in binary mixtures.

By definition,

$$\mathbf{J}_A + \mathbf{J}_B = 0 \tag{1.12}$$

Again, an equation relating the diffusive fluxes to the driving forces that cause diffusion is required: Fick's law is a good empirical relation for many binary systems. For component B, this equation is written as

$$\mathbf{J}_B = -c\mathscr{D}_{BA} \nabla x_B \tag{1.13}$$

---

[2] Average of the velocity of the individual molecules.

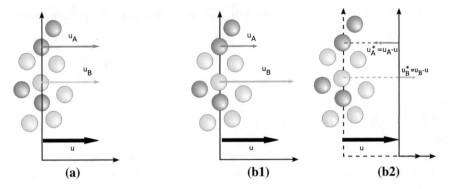

Fig. 1.2 Convection and diffusion processes in a binary mixture: **a** convection: all the components move with the same average velocity; **b** diffusion: components A and B move with different velocity; **b1** velocities are measured with respect to a fixed frame; **b2** diffusion velocities are measured with respected to a frame in motion with the mean velocity of the mixture

Considering Eq. (1.12), it is easy to prove that $\mathscr{D}_{AB} = \mathscr{D}_{BA}$, i.e., there is just one diffusivity for a binary mixture.

1.3   Diffusivity

Fick's law defines diffusivity of a binary mixture $\mathscr{D}_{AB}$ as the proportionality constant between diffusive flux and molar fraction (or concentration) gradient. Values of $\mathscr{D}_{AB}$ can be obtained by experimental measurements.

Some values of diffusivity in gases, liquids, or tissues that are interesting for biomedical applications are reported in Tables 1.1 and 1.2. It is worth noting that diffusivity in liquids are four orders of magnitude lower than diffusivity in gases at atmospheric pressure; such a low diffusivity means that diffusion in liquids is a slow process, often limiting the overall kinetics.

Table 1.1   Diffusivities in gases and liquids

	In air (273 K, P = 1 atm) cm²/s	In water (298 K) $\times 10^5$ cm²/s
CO_2	0.138	1.96
N_2		1.9
O_2	0.178	2.5
H_2O	0.220	
Benzene	0.077	
Ethanol		1.28
Glycerin		0.94
Glucose		0.9

Table 1.2 Diffusivities of gases in biological tissues (data from Lango et al. [2])

	Tissue water content (%)	O_2, cm^2/s	N_2, cm^2/s
Water	1	3.2	
Plasma	0.93	2.54	2.17
Brain	0.81	1.3	
Aorta, tunics media	0.73	0.965	0.918
Aorta, tunica intima	0.69	0.837	0.782
Fat tissue	0.47	0.291	0.15

Diffusivities are low for large molecules; more specifically, the diffusivity of a dilute permeant in a liquid solvent decreases with increasing permeant size and solvent viscosity, as described by Stokes–Einstein equation

$$\mathscr{D}_{AB} = \frac{k_B T}{6\pi \mu_B R_A} \tag{1.14}$$

where k_B is the Boltzmann's constant, μ_B the solvent viscosity, and R_A the solute molecule radius. The effect of the permeant size is even stronger in polymer or complex system as the intracellular space (see Fig. 1.3).

Finally, it is worth considering how diffusivity varies with temperature and pressure. As for gases, diffusivity increases rapidly with temperature (proportionally to the absolute temperature raised to the power of 1.5–1.75) and is inversely proportional to pressure. As for liquids, Eq. 1.14 suggests that $\mathscr{D}_{AB}\mu_B/T$ must be constant. In many cases, a significant dependence of the diffusivity on the liquid mixture composition is observed; values reported in the literature usually refer to diffusivity at infinite dilution of the permeant.

Fig. 1.3 Diffusivities of different compounds in water and in intracellular space (data from Lightfoot and Duca [3])

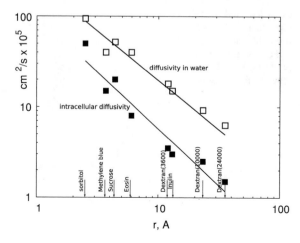

1.4 Local Analysis of Mass Transport Phenomena and Evaluation of the Concentration Profiles

Diffusion problems, i.e., the evaluation of the concentration profile in a system where diffusion processes occur, can be solved imposing the mass balance for each chemical species over an infinitesimal control volume, using phenomenological laws for diffusive fluxes (Fick's law) and for chemical reaction rates.

Here, the general formulation of local mass balance equations is reported, both in steady-state and unsteady-state conditions; the solutions are considered in the simplest case of zero velocity (no significant convection) and without any chemical reaction. In the following sections, more complex conditions are discussed.

For sake of simplicity, let us first consider an unidimensional system, i.e., a system where everything varies only along one spatial variable (z_x direction in Fig. 1.4) and a fixed control volume $\Delta x \, \Delta y \, \Delta z$: in this case, the mass balance equation may be written as

$$\frac{\partial}{\partial t}(c_A \Delta x \Delta y \Delta z) = N_{Ax} \Delta y \Delta z|_x - N_{Ax} \Delta y \Delta z|_{x+\Delta x} + r_A \Delta x \Delta y \Delta z \qquad (1.15)$$

where the term on the l.h.s represents the rate of increase of A moles in the control volume, while the first two terms on the r.h.s. account for the rate of A in and out the control volume, through the surface at x and $x + \Delta x$; finally, the last term accounts for the net rate of production of A in the control volume, due to chemical reactions with a reaction rate per unit volume r_A. By dividing all the expression for $\Delta x \Delta y \Delta z$ and taking the limit as $\Delta x \to 0$, we get

$$\frac{\partial c_A}{\partial t} = -\frac{\partial N_{Ax}}{\partial x} + r_A \qquad (1.16)$$

Fig. 1.4 Fluid flowing through a fixed volume element; a flow rate through the surface at x and $x + \Delta x$ and chemical reaction that produces A inside the volume are considered

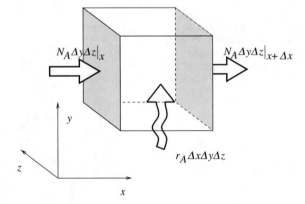

If the flux of A has nonzero components in y and z directions, mass balance of A is expressed in a general form as

$$\frac{\partial c_A}{\partial t} = -\frac{\partial N_{Ax}}{\partial x} - \frac{\partial N_{Ay}}{\partial y} - \frac{\partial N_{Az}}{\partial z} + r_A \tag{1.17}$$

or, using vector notation[3]:

$$\frac{\partial c_A}{\partial t} = -\nabla \cdot \mathbf{N}_A + r_A \tag{1.18}$$

The above equation may be written in any system of coordinate, providing the right expression for $\nabla \cdot \mathbf{N}_A$.

Let us now refer to a zero velocity system, where $\mathbf{N}_A = \mathbf{J}_A$ and the diffusive fluxes are given by the Fick's law; if no reaction occurs, for a constant molar density system (dilute solution or low pressure gas) we get

$$\frac{\partial c_A}{\partial t} = -\mathscr{D}_{AB}\left[\frac{\partial^2 c_A}{\partial x^2} + \frac{\partial^2 c_A}{\partial y^2} + \frac{\partial^2 c_A}{\partial z^2}\right] \tag{1.19}$$

or, using vector notation[4]:

$$\frac{\partial c_A}{\partial t} = -\mathscr{D}_{AB}\nabla^2 c_A \tag{1.20}$$

Even if Eq. (1.20) has been derived referring to a rectangular volume element, it is a general equation (*equation of continuity of the component A*) which describes the rate of change of the concentration of a non-reacting component A at a fixed point of space. Actually, in order to solve a diffusion problems we can set up a mass balance of component A over an infinitesimal control volume and use Fick's law for diffusive flux, or we can use straightforwardly the equation of continuity (1.20).

The equation of continuity of A (1.20) is a partial differential equation of first order with respect to time and of second order with respect to the spatial coordinates: therefore, suitable initial and boundary conditions are required. While initial condition is simply given by the concentration profile at $t = 0$, some attention is required for the boundary conditions. These can be given

---

[3]$\nabla \cdot \mathbf{v}$ is the *divergence* of the vector $\mathbf{v}$; in rectangular coordinates

$$\nabla \cdot \mathbf{v} = \frac{\partial v_x}{\partial x} + \frac{\partial v_y}{\partial y} + \frac{\partial v_z}{\partial z}$$

For other coordinate systems, see Appendix A at the end of this chapter.

[4]$\nabla^2 s$ is called Laplacian of a scalar s and it is defined as $\nabla \cdot \nabla s$. In rectangular coordinates,

$$\nabla^2 s = \frac{\partial^2 s}{\partial x^2} + \frac{\partial^2 s}{\partial y^2} + \frac{\partial^2 s}{\partial z^2}$$

For other coordinates, see Appendix A at the end of this chapter.

Fig. 1.5 Boundary
conditions for mass transport
in a fluid phase

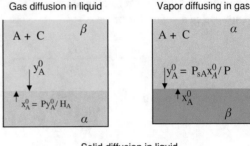

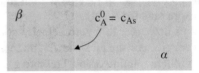

- by specifying the A flux through some surface (Newmann's condition): e.g., $J_{A,x} = 0$ (and therefore $\partial c_A/\partial x = 0$) at $x = 0$ where an impermeable surface is present;
- by specifying the A concentration at some surface (Dirichlet's condition): e.g., $c_A = c_A^0$ at $x = 0$. In many cases, the surface considered is the interface between the phase α where diffusion occurs and a different phase, β; in all the situations, it is assumed that the α- and β-phase concentrations at the interface are in thermodynamic equilibrium

$$c_A^\alpha\big|_{x=0} = f\left(c_A^\beta\big|_{x=0}\right) \tag{1.21}$$

Relations to describe the equilibrium conditions are regarded as known and depend on the nature of the two phases.

Some examples of boundary conditions are reported in Fig. 1.5.

1.5   Diffusion Characteristic Time

Let us consider the unsteady-state diffusion of A in a medium that initially does not contain A; at time $t = 0$, the concentration of A on the surface at $x = 0$ suddenly increases to c_A^0. The increase produces a time-dependent concentration profile that penetrates into the slab. Time dependence of the concentration profile can be obtained by solving the equation:

$$\frac{\partial c_A}{\partial t} = -\mathscr{D}_{AB}\frac{\partial^2 c_A}{\partial x^2} \tag{1.22}$$

with suitable initial and boundary conditions:

$$t < 0, \forall x \qquad c_A = 0 \tag{1.23}$$

$$t \geq 0, \ x = 0 \qquad c_A = c_A^0 \tag{1.24}$$

It is interesting to rewrite the above equation in terms of dimensionless variables:

$$\tilde{c}_A = c_A / c_A^0 \qquad \tilde{x} = x / \ell \tag{1.25}$$

where ℓ is the spatial characteristic dimension of the problem. In terms of dimensionless variables, mass balance equation is rewritten as

$$\frac{\partial \tilde{c}_A}{\partial t} = -\frac{\mathscr{D}_{AB}}{\ell^2} \frac{\partial^2 \tilde{c}_A}{\partial \tilde{x}^2} \tag{1.26}$$

$$t < 0, \forall \tilde{x} \quad \tilde{c}_A = 0 \qquad t \geq 0, \ \tilde{x} = 0 \quad \tilde{c}_A = 1 \tag{1.27}$$

The ratio $\ell^2 / \mathscr{D}_{AB}$ has the dimension of a time, and it is known as *diffusion characteristic time*. It physically represents the time required for the concentration c_A^0 to penetrate to a depth ℓ; on the other hand, it can be stated that in a time t_D, the concentration profile penetrates a slab of depth $\sqrt{\mathscr{D}_{AB} t_D}$.

1.6 Diffusion and Chemical Reaction

Diffusion and chemical reaction often occur together in biological and biomedical systems: for example, oxygen diffusion in metabolizing tissue, nutrients and oxygen transport in bioreactors, enzymatic biosensors measuring the concentration of different analytes (e.g., glucose concentration in blood) can be considered. In these systems, chemical reaction kinetics and diffusion rate interact to determine the overall kinetics of the process and, depending on the working conditions, chemical reaction, or diffusion can control the overall kinetics.

In order to model the processes where diffusion and chemical reaction occur, it is fundamental to distinguish between heterogeneous reactions, which occur on a surface (a catalyst surface, a surface where an enzyme is immobilized) and homogeneous reactions, which occur within the solution volume. A schematic plot representing the two different reaction types is reported in Fig. 1.6. In the first case (heterogeneous reaction), reactant has to diffuse through a layer to reach the surface where reaction occurs. The reaction product then diffuses toward the bulk of the system; reaction rate is proportional to the surface area and it is determined by the reactant concentration on the surface reaction. The diffusion process is "in series" with the reaction process. As for homogeneous reaction, the reactant diffuses in the fluid phase and, as it diffuses, it also participates in a chemical reaction, with a reaction rate per unit

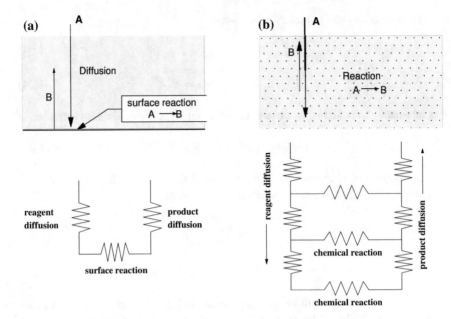

Fig. 1.6 Diffusion and reaction: **a** in series; **b** in parallel

volume that, at any point, depends on the local concentration of the reactant; the diffusion and the reaction processes occur "in parallel."

1.6.1 Diffusion and Reaction in Series

Let us consider the system depicted in Fig. 1.7. In this case, no reaction occurs in the film, so steady-state mass balance equation results in

$$\frac{\mathrm{d}^2 c_A}{\mathrm{d}x^2} = 0 \tag{1.28}$$

Fig. 1.7 Diffusion and heterogeneous reaction

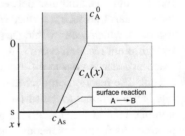

that is, we have a linear concentration profile in the film. Boundary condition at $x = 0$ is $c_A = c_A^0$, while at $x = s$ the rate at which A reaches the surface must equal the rate at which it is consumed by the chemical reaction (i.e., the flux of A at $x = s$ must be equal to the reaction rate per unit surface area). Referring to a first-order reaction, with a reaction rate proportional to the A concentration on the reaction surface, the second boundary condition is written as

$$x = s \qquad -\mathscr{D}_{AB}\frac{dc_A}{dx} = r'_A\big|_{x=s} = k'\,c_A\big|_{x=s} \tag{1.29}$$

It is easily derived that the concentration profile is given by

$$\frac{c_A}{c_{A0}} = 1 - \frac{\dfrac{k's}{\mathscr{D}_{AB}}}{\dfrac{k's}{\mathscr{D}_{AB}}+1}\,\frac{x}{s} \tag{1.30}$$

with a surface concentration

$$c_{As} = \frac{1}{1+\dfrac{k's}{\mathscr{D}_{AB}}}c_A^0 < c_A^0 \tag{1.31}$$

The effective reaction rate per unit surface area is then

$$r_A = \frac{1}{1+\dfrac{k's}{\mathscr{D}_{AB}}}k'c_A^0 \tag{1.32}$$

The conversion of A proceeds at a rate lower than the theoretical rate $k'c_A^0$, obtained if the surface would be exposed to the bulk concentration of A; the finite rate of the diffusion process reduces the rate of the process and the higher is the ratio $k's/\mathscr{D}_{AB}$, the lower is the overall reaction rate. Two limiting cases can be considered:

- if $k's/\mathscr{D}_{AB} \ll 1$, $c_{As} \approx c_A^0$ and $r'_A \approx k'c_A^0$: the process is under kinetic control; an increase of the reaction rate (e.g., with an increase of the catalyst concentration on the surface) results in an equal increase of the overall process rate; a reduction of the diffusional resistance (e.g., a reduction of the thickness of the diffusion layer) does not affect the overall kinetics;
- if $k's/\mathscr{D}_{AB} \gg 1$, $c_{As} \approx 0$ and $r'_A \approx \mathscr{D}_{AB}c_A^0/s$: the process is under diffusion control; an increase in the reaction rate is completely ineffective on the overall process kinetics; on the contrary, a reduction of diffusional resistance results in an increase of the overall rate.

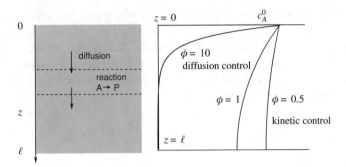

Fig. 1.8 Diffusion and homogeneous reaction in parallel

1.6.2 Diffusion and Reaction in Parallel

Here, we consider a slab where the component A diffuses and participates in a homogeneous chemical reaction, with a reaction rate per unit volume that depends on the concentration of A (see Fig. 1.8); for a first-order reaction $r_A = kc_A$. Mass balance equation in steady state must account for A consumption due to the reaction rate[5]:

$$\mathscr{D}_{AB}\frac{d^2c_A}{dx^2} - r_A = \mathscr{D}_{AB}\frac{d^2c_A}{dx^2} - kc_A = 0 \tag{1.33}$$

Boundary condition are given by: $x = 0$ $c_A = c_A^0$ and $x = s$ $dc_A/dx = 0$; the latter boundary condition states that there is no flux of A through the surface at $x = s$.

Before solving Eq. (1.33), it is convenient to rewrite it in dimensionless form, by defining:

$$\tilde{c}_A = \frac{c_A}{c_A^0} \qquad \tilde{x} = \frac{x}{\ell} \tag{1.34}$$

We obtain:

$$\frac{d^2\tilde{c}_A}{d\tilde{x}^2} - \frac{k\ell^2}{\mathscr{D}_{AB}}\tilde{c}_A = 0 \tag{1.35}$$

with the boundary conditions:

$$\tilde{x} = 0 \quad \tilde{c}_A = 1; \qquad \tilde{x} = 1 \quad \frac{d\tilde{c}_A}{d\tilde{x}} = 0 \tag{1.36}$$

---

[5]Equation of continuity for a component A which undergoes a chemical reaction is written as

$$\frac{\partial c_A}{\partial t} = \mathscr{D}_{AB}\nabla^2 c_A - r_A.$$

The above equation defines a dimensionless group, usually known as Thiele's modulus:

$$\phi = \ell \sqrt{\frac{k}{\mathcal{D}_{AB}}} \tag{1.37}$$

It is important to point out that:

- if $\phi \to 0$, $\tilde{c}_A \sim 1$ everywhere; in this condition, diffusion of A is fast enough so that there is no decrease in A concentration in the slab; as a consequence, for a slab of unit area the overall consumption rate is $k c_A^0 \ell$ and the process is controlled by the reaction kinetics;
- if $\phi \gg 1$, $\tilde{c}_A \sim 0$ almost everywhere, except near to the interface at $x = 0$; in other words, since diffusion is slow, c_A drops from c_A^0 to 0 in a thin layer near the surface at $x = 0$ with width $\sim 1/\phi$. As a consequence, for a slab of unit area, the overall consumption rate of A can be evaluated as the flux of A entering the slab through the surface at $x = 0$ (at steady state, all A entering the slab is consumed inside the slab, since there is no flux of A through the surface at ℓ) and it is given by $\mathcal{D}_{AB} c_A^0 \phi = \sqrt{k \mathcal{D}_{AB}} c_A^0 \ell$. Usually, in such a condition the process is referred as diffusion controlled, even if both the diffusion and the reaction rate affects the overall kinetics.

The general solution of Eq. 1.35 gives the concentration profile inside the slab (see Fig. 1.8):

$$\tilde{c}_A = \frac{\cosh \left[\phi \left(1 - \tilde{x} \right) \right]}{\cosh \phi} \tag{1.38}$$

and the consumption of A within the slab is given by $k c_A^0 \ell \tanh \phi / \phi$. Actually, the limited diffusion rate reduces the overall reaction rate with respect to the theoretical reaction rate corresponding to $c_A = c_A^0$ throughout the slab; therefore, the term $\tanh \phi / \phi$ has the physical meaning of an efficiency factor.

1.6.3 Oxygen Transport to Tissue

Oxygen has to be continuously supplied to living tissues where it is consumed by metabolic processes; in fact, an insufficient oxygen supply results in a reduction of the oxygen concentration in the tissue and its anoxia-induced death.

In order to describe oxygen transport in living tissue, Krogh's cylinder model has been proposed: an arrangement of nearly parallel, uniformly spaced capillaries, of radius R_1, each of them responsible for the oxygen delivery to a tissue region of radius R_2 (see Fig. 1.9).

Let us consider the oxygen transport in the tissue at steady state, accounting for an oxygen consumption rate per unit volume $\mathcal{G}$:

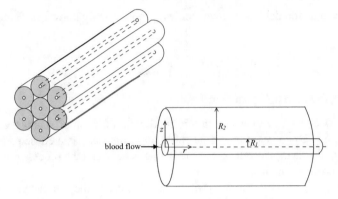

Fig. 1.9 Schema of a living tissue oxygenation according to the Krogh's model

$$\mathscr{D}_T \left[\frac{1}{r} \frac{\partial}{\partial r} \left(r \frac{\partial c}{\partial r} \right) + \frac{\partial^2 c}{\partial z^2} \right] + \mathscr{G} = 0 \qquad (1.39)$$

where $\mathscr{D}_T$ is the oxygen diffusivity in the tissue and the term in square brackets is $\nabla^2 c$ in cylindrical coordinates. In the above equation, the axial diffusion term ($\partial^2 c / \partial z^2$) is negligible; furthermore, the oxygen concentration in the tissue may be related to the oxygen partial pressure of a gas phase in equilibrium with the tissue (usually simply referred as oxygen pressure in the tissue) by considering the oxygen Henry's constant in tissue, H_T, defined as follows:

$$H_T = \frac{dp}{dc} \qquad (1.40)$$

Therefore, we may write

$$\frac{1}{r} \frac{\partial}{\partial r} \left(r \frac{\partial p}{\partial r} \right) - \frac{\mathscr{G} H_T}{\mathscr{D}_T} = 0 \qquad (1.41)$$

that must be solved with the boundary conditions

$$r = R_1 \quad p = p_B; \qquad r = R_2 \quad \frac{\partial p}{\partial r} = 0 \qquad (1.42)$$

The first one states that on the capillary the tissue is in equilibrium with the blood with an oxygen pressure p_B; the second one states that, for symmetry, there is no flux through the surface at $r = R_2$.[6] Integrating with $\tilde{R} = R_2/R_1$ and $\alpha = \mathscr{G} H_T / D_T$, we get:

---

[6]This condition assumes $p > 0$ at $r = R_2$. In the presence of an anoxic region, this boundary condition must be modified.

$$p(r) - p_B = -\frac{\mathcal{G} H_T}{4 \mathcal{D}_T} R_1^2 \left[\left(\tilde{R} \right)^2 \ln \left(\frac{r}{R_1} \right)^2 - \left(\frac{r}{R_1} \right)^2 + 1 \right] \tag{1.43}$$

and the minimum oxygen pressure at $r = R_2$:

$$p(R_2) - p_B = -\frac{\mathcal{G} H_T}{4 \mathcal{D}_T} R_1^2 \left[\left(\tilde{R} \right)^2 \ln \left(\tilde{R} \right)^2 - \left(\tilde{R} \right)^2 + 1 \right] \tag{1.44}$$

It is worth noting that the oxygen pressure in the tissue depends not only on the oxygen consumption rate, but also on its diffusivity in the tissue.

The flux of oxygen transferred from blood to tissue is given by

$$N_{O_2} = -\frac{\mathcal{D}_T}{H_T} \left. \frac{\partial p}{\partial r} \right|_{r=R_1} = \frac{\mathcal{G} R_1}{2} \left(\tilde{R}^2 - 1 \right) \tag{1.45}$$

It is worth noting that the oxygen flux does not depend on the oxygen pressure in the blood and it is constant along the capillary. As a consequence, the oxygen pressure in the blood varies linearly according to the oxygen balance:

$$\frac{\pi R_1^2 v}{H_B} \frac{\partial p_B}{\partial z} = N_{O_2} 2\pi R_1 \tag{1.46}$$

where H_B is the Henry's constant for oxygen in blood ($H_B = \mathrm{d}p_B/\mathrm{d}c_B$) and v is the blood velocity; by substituting Eq. (1.45) and integrating along the capillary, starting from p_B equal to the arterial oxygen pressure, p_a, at $z = 0$, we get

$$p_B = p_a - \frac{\mathcal{G} H_B}{v} \left(\tilde{R}^2 - 1 \right) z \tag{1.47}$$

The oxygen pressure profile along a capillary and in the surrounding tissue is reported in Fig. 1.10. The oxygen pressure at the venous end of the capillary depends on the oxygen consumption in the tissue and on the blood flow rate in the capillary; an increase in the oxygen demand by the tissue or a reduction on the blood flow rate result in a decrease of venous oxygen pressure; as we have seen before, the oxygen in the tissue is still lower than in blood and the minimal oxygen pressure obtained for $z = L$ and $r = R_2$ is given by

$$p_{min} = p_a - \frac{\mathcal{G} H_B L}{v} \left\{ \left(\tilde{R}^2 - 1 \right) + \frac{(H_T/H_B) v}{4 \mathcal{D}_T L} R_1^2 \left[\left(\tilde{R} \right)^2 \ln \left(\tilde{R} \right)^2 - \left(\tilde{R} \right)^2 + 1 \right] \right\} \tag{1.48}$$

If this pressure drops under some minimal value, local anoxia and tissue death can occur.

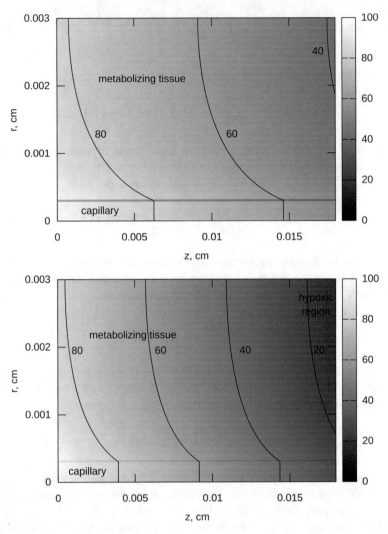

Fig. 1.10 Oxygen pressure profile along a capillary and in the surrounding tissue. The following representative parameters are used: $R_1 = 3\,\mu m$, $R_2 = 30\,\mu m$, $L = 180\,\mu m$, $\mathscr{G} = 3.72 \times 10^{-8}\,mol/(s\,cm^3)$, $H_T = 7.4 \times 10^8\,mmHg/(mol/cm^3)$, $\mathscr{D}_T = 1.7 \times 10^5\,cm^2/s$, $H_B = 2.6 \times 10^7\,mmHg/(mol/cm^3)$. In the upper figure, a normal blood flow rate ($v = 400\,\mu m/s$) is considered; in the bottom the effect of a reduction of the blood flow rate ($v = 250\,\mu m/s$) is considered

1.7 Diffusion and Convection

In Sect. 1.4 the general formulation of mass balance equation has been considered; in particular, Eq. (1.18) reports the mass balance equation of component A in term of the A fluxes at any point of the system. It is worth noting that the fluxes here considered are the total diffusive+convective fluxes and only for zero velocity systems considered until now we have $\mathbf{N}_A = \mathbf{J}_A$.

Let us turn to consider systems where external forces[7] cause a bulk fluid motion with $\mathbf{u}$ velocity, so that a convective motion superimposes to diffusion; referring to diluted solutions the flux of A is given by

$$\mathbf{N}_A = \mathbf{J}_A + c_A\mathbf{u} = -\mathscr{D}_{AB}\nabla c_A + c_A\mathbf{u} \tag{1.49}$$

Obviously, both $\mathbf{J}_A$ and $\mathbf{u}$ are vectors, so they sum up accounting for their directions; in particular, if $\mathbf{J}_A$ and $\mathbf{u}$ have the same direction, their magnitude sum up and convection and diffusion reinforce each other for A transport, while, if they have opposite direction a net flux that is the difference between them occurs. The first case applies for convection enhanced diffusion; the second one occurs, for example, in membrane polarization. In the following, we discuss the general feature of diffusion and convection referring to the case of convection enhanced diffusion.

Let us consider two surface at $z = 0$ and $z = L$; the concentrations of A on the two surfaces are c_{A0} and $c_{AL} < c_{A0}$, respectively, and fluid moves in the z direction with velocity u_z (see Fig. 1.11): it is evident that both diffusion and convection push the component A toward $z = L$. Mass balance equation in a control volume between z and $z + \Delta z$ results in[8]:

$$uc_A|_z - \mathscr{D}_A \left.\frac{dc_A}{dz}\right|_z = uc_A|_{z+\Delta z} - \mathscr{D}_A \left.\frac{dc_A}{dz}\right|_{z+\Delta z} \tag{1.50}$$

or

$$u\frac{dc_A}{dz} - \mathscr{D}_A \frac{dc_A^2}{dz^2} = 0 \tag{1.51}$$

that must be solved with the boundary conditions $z = 0$ $c_A = c_{A0}$ and $z = L$ $c_A = c_{AL}$. In order to get a general results, the above equations are rewritten in terms of dimensionless variables defined as

---

[7]Bulk fluid motion can be generated by diffusion itself without external driving forces; since this case is not interesting for artificial organ design, it is not considered here.

[8]More generally, the continuity equation in the presence of nonzero fluid velocity is written as

$$\mathbf{u} \cdot \nabla c_A = \nabla \cdot (\mathscr{D}_A \nabla c_A)$$

or for constant $\mathscr{D}_A$,

$$\mathbf{u} \cdot \nabla c_A = \mathscr{D}_A \nabla^2 c_A.$$

Fig. 1.11 Concentration profiles for a monodimensional convection-diffusion problem

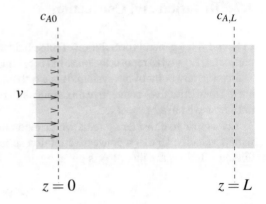

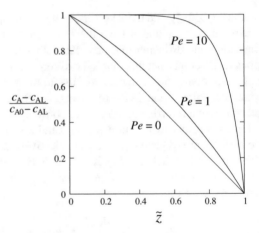

$$\tilde{c} = \frac{c_A - c_{AL}}{c_{A0} - c_{AL}} \qquad \tilde{z} = \frac{z}{L} \tag{1.52}$$

We get:

$$\frac{d^2\tilde{c}}{d\tilde{z}^2} - \frac{uL}{\mathscr{D}_{AB}}\frac{d\tilde{c}}{d\tilde{z}} = 0 \qquad \tilde{z} = 0 \ \tilde{c} = 1; \quad \tilde{z} = 1 \ \tilde{c} = 0 \tag{1.53}$$

The above equation introduces a dimensionless group, usually named as Peclet number:

$$Pe = \frac{uL}{\mathscr{D}_{AB}} \tag{1.54}$$

It can be easily proved that Pe represents the ratio between the diffusion time $L^2/\mathscr{D}_{AB}$ (time required to diffuse through a length L) and the convection time L/u (time required to move to L at velocity u): the larger the Pe, the higher the convective contribution to the A transport, while $Pe \ll 1$ corresponds to negligible convection effect and A is simply transported by diffusion.

Solution of Eq. (1.53) gives:

$$\tilde{c} = \frac{e^{Pe} - e^{Pe\tilde{z}}}{e^{Pe} - 1} \qquad (1.55)$$

that simplifies in $\tilde{c} = 1 - \tilde{z}$ (i.e., the linear concentration profile obtained for zero velocity diffusion) for low Pe. The flux of A is given by

$$
\begin{aligned}
N_{Az} &= -\frac{\mathscr{D}_{AB}\,(c_{A0} - c_{AL})}{L}\frac{\mathrm{d}\tilde{c}}{\mathrm{d}\tilde{z}} + v\left[\tilde{c}\,(c_{A0} - c_{AL}) + c_{A0}\right] \\
&= \frac{(c_{A0} - c_{AL})\,\mathscr{D}_{AB}}{L}Pe\left[\frac{1}{e^{Pe} - 1} + \frac{c_{A0}}{(c_{A0} - c_{AL})}\right] \\
&= uc_{A0}\left[1 + \frac{c_{A0} - c_{AL}}{c_{A0}}\frac{1}{e^{Pe} - 1}\right]
\end{aligned}
\qquad (1.56)
$$

Symbols

c	Total molar concentration
c_i	Molar concentration of the component i
$\mathscr{D}_{ij}$	Diffusion coefficient of the binary mixture of the components i and j
$\mathscr{D}_T$	Diffusivity in a biological tissue
$\mathscr{G}$	Consumption rate per unit volume
H_B	Henry's constant in blood
H_T	Henry's constant in a tissue
$J_{i,x}$	Diffusive flux of the component i in the x direction
$\mathbf{J}_i$	Diffusive flux (vector) of the component i
k_B	Boltzman constant
ℓ	Characteristic length
L	Length
$\mathbf{N}_i$	Flux (vector) of the component A with respect to a fixed frame
p_i	Partial pressure of the component i
Pe	Peclet number
R	Particle, molecule, sphere, or pipe radius
r_i	Rate of reaction of the component i per unit volume
r_i'	Rate of reaction of the component i per unit surface area
T	Temperature
$\mathbf{u}$	Velocity (vector)
$\mathbf{u}_A$	Velocity (vector) of component A
u_x	Velocity in the direction x
$\mathbf{u}^*$	Diffusion velocity (vector)
v	Blood velocity in a capillary
x_A	Molar fraction of component A
μ	Viscosity
ϕ	Thiele's modulus

Subscript

A, B   With respect to the component A, B
r With respect to the radial direction
x, y, z With respect to the direction x, y, and z

Superscript

α, β Phases
~   Dimensionless variable

Appendix A: Vector Differential Operators

A.1 Cartesian Coordinates

Gradient	∇s	$[\nabla s]_x = \frac{\partial s}{\partial x}$ $[\nabla s]_y = \frac{\partial s}{\partial y}$ $[\nabla s]_z = \frac{\partial s}{\partial z}$
Laplacian	$\nabla^2 s$	$\nabla^2 s = \frac{\partial^2 s}{\partial x^2} + \frac{\partial^2 s}{\partial y^2} + \frac{\partial^2 s}{\partial z^2}$
Divergence	$\nabla \cdot \mathbf{v}$	$\nabla \cdot \mathbf{v} = \frac{\partial v_x}{\partial x} + \frac{\partial v_y}{\partial y} + \frac{\partial v_z}{\partial z}$

A.2 Cylindrical Coordinates

Gradient	∇s	$[\nabla s]_r = \frac{\partial s}{\partial r}$ $[\nabla s]_\theta = \frac{1}{r}\frac{\partial s}{\partial \theta}$	$[\nabla s]_z = \frac{\partial s}{\partial z}$
Laplacian	$\nabla^2 s$	$\nabla^2 s = \frac{1}{r}\frac{\partial}{\partial r}\left(r \frac{\partial s}{\partial r}\right) + \frac{1}{r^2}\frac{\partial^2 s}{\partial \theta^2} + \frac{\partial^2 s}{\partial z^2}$	
Divergence	$\nabla \cdot \mathbf{v}$	$\nabla \cdot \mathbf{v} = \frac{1}{r}\frac{\partial}{\partial r}(r v_r) + \frac{1}{r}\frac{\partial v_\theta}{\partial \theta} + \frac{\partial v_z}{\partial z}$	

A.3 Spherical Coordinates

Gradient	∇s	$[\nabla s]_r = \frac{\partial s}{\partial r}$ $[\nabla s]_\theta = \frac{1}{r}\frac{\partial s}{\partial \theta}$	$[\nabla s]_\phi = \frac{1}{r \sin\theta}\frac{\partial s}{\partial \phi}$
Laplacian	$\nabla^2 s$	$\nabla^2 s = \frac{1}{r^2}\frac{\partial}{\partial r}\left(r^2 \frac{\partial s}{\partial r}\right) + \frac{1}{r^2 \sin\theta}\frac{\partial}{\partial \theta}\left(\sin\theta \frac{\partial s}{\partial \theta}\right) + \frac{1}{r^2 \sin^2\theta}\frac{\partial^2 s}{\partial \phi^2}$	
Divergence	$\nabla \cdot \mathbf{v}$	$\nabla \cdot \mathbf{v} = \frac{1}{r^2}\frac{\partial}{\partial r}(r^2 v_r) + \frac{1}{r \sin\theta}\frac{\partial}{\partial \theta}(v_\theta \sin\theta) + \frac{1}{r \sin\theta}\frac{\partial v_\phi}{\partial \phi}$	

Chapter 2
Mass Transfer Coefficient

2.1 Introduction

The analysis reported in the previous chapter allows to describe the concentration profile and the mass fluxes of components in a mixture by solving local mass balance equations, once the mixture velocity field is known. Such analysis allows the concentration profiles to be evaluated in solids (no velocity) or fluids in laminar flow, when the velocity can be determined via the motion equation (Navier–Stokes equations); the only difficulties in this approach are linked to the mathematical and numerical complexity, which can be important, especially for unsteady state and/or 3D problems. A completely different scenario appears when we deal with turbulent flow, in which elements of fluid do not flow in an orderly manner, one layer over another, but in a chaotic mode, with rapid velocity fluctuations and nonzero instantaneous velocity components, even in the direction perpendicular to the bulk fluid motion. As an example, in the case of turbulent fluid flow past a flat solid surface, eddies with a size several orders of magnitude larger than the mean molecular free path arise and the velocity field shows significant instantaneous components also perpendicularly to the surface; such phenomena contribute significantly to the mass transport process and must be accounted for. If mass transfer between the liquid and the surface occurs in these conditions, thanks to turbulent mixing, the fluid composition varies only slightly with the position in the bulk of fluid, whereas a significant composition gradient may be observed in a thin layer of fluid near the surface.

Considering that:

- even though turbulence is a widely studied topic, we are far from a satisfactory understanding of the turbulent flow mechanism and, as a consequence, of mass transport in turbulent flow
- the mixture composition in the bulk of the fluid in turbulent flow is fairly uniform
- in many relevant engineering applications, a detailed quantitative evaluation of the concentration profile is not required

© Springer-Verlag London 2017
M.C. Annesini et al., *Artificial Organ Engineering*,
DOI 10.1007/978-1-4471-6443-2_2

we turn our attention to lumped-parameter models, aimed at evaluating the mean concentration of the system (in the bulk) and, in some cases, its evolution with time. This simplified analysis requires only the knowledge of the mass fluxes exchanged by the system with the surroundings, through its boundary surface.

Mass transfer coefficients are defined with the aim of evaluating these fluxes in a simple way, with a wise use of theoretical analysis and empirical results.

2.2 Definition of Mass Transport Coefficients

Mass transport coefficient is defined so that

diffusive flux = (mass transport coefficient) × (concentration difference)

or

$$k_c = \frac{J_A|_s}{c_A^0 - c_{Af}} \tag{2.1}$$

where $J_A|_s$ is the diffusive flux of A on the surface between the system and the surrounding, c_A^0 and c_{Af} are the concentrations of A on the surface and on the bulk of fluid, respectively.[1]

Sometimes, the difference in molar fractions or, for a gas phase, in partial pressures between the surface and the bulk of fluid can be considered as the driving force for mass transport; as a consequence, different mass transport coefficient can be also defined as follows:

$$k_x = \frac{J_A|_s}{x_A^0 - x_{Af}} \qquad k_p = \frac{J_A|_s}{p_A^0 - p_{Af}} \tag{2.3}$$

Since

$$k_c \left(c_A^0 - c_{Af}\right) = k_x \left(x_A^0 - x_{Af}\right) = k_p \left(p_A^0 - p_{Af}\right) \tag{2.4}$$

it can be easily proved that

$$k_x = k_c c \qquad k_p = \frac{k_c}{RT} \tag{2.5}$$

---

[1]The bulk concentration is the concentration in the fluid far from the surface or, for flow in conduits, the cup-mixing concentration defined as

$$c_{Af} = \frac{\int_S v c_A \, dS}{\int_S v \, dS} \tag{2.2}$$

so that the flow rate of A through the duct section can be given as $c_{Af} \int_S v \, dS$.

The mass transfer coefficients so defined must account for physical properties of the system (in particular, the diffusivity, $\mathcal{D}_{AB}$, fluid density, ρ, and viscosity, μ), and for the fluid motion.

2.3 Evaluation of Mass Transport Coefficient

It is evident that, if the mass transport coefficient is known, the evaluation of the flux of A is straightforward, while all the procedure described in the previous section is useless if a method to calculate the mass transport coefficient is not available. Obviously, it is possible to get mass transport coefficient from the results of mass transport experiments, but what we do really need are some correlations of the experimental results that allow to calculate the mass transport coefficient in a variety of conditions. Theoretical considerations, less or more complex, are of paramount utility to wisely build these correlations.

The oldest and most obvious approach is based on the so-called *film model* (see Fig. 2.1). In such model, the entire concentration difference between the bulk of fluid and the surface is considered to be localized in a viscous thin layer (film); within the film, which is adjacent to the surface, solute transport occurs by molecular diffusion, so that the flux of component A is given by:

$$J_A = k_c \left(c_A^0 - c_{Af} \right) = \frac{\mathcal{D}_{AB}}{\delta} \left(c_A^0 - c_{Af} \right) \tag{2.6}$$

where δ is the film thickness, which is assumed to depend on fluid motion conditions.

Fig. 2.1 Concentration profile near the interface in the film model

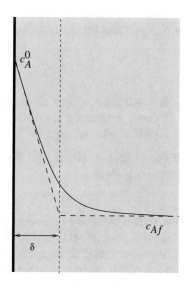

The film model does not allow to predict the mass transfer coefficient, since we are unable to evaluate δ, but it predicts that mass transport coefficients for different solutes, under the same fluid flow conditions, are in the same ratio of their molecular diffusivities. Experimental data actually show that mass transport coefficients of different component in the same solvent and in the same fluid dynamic conditions increase with diffusivities to a power ranging from $1/2$ to $1/3$, and, therefore, more complex models must be considered to get fairly good correlations.

More reliable results can be obtained from rigorous dimensional analysis, boundary layer theory or penetration theory: since a deep analysis of mass transport models is out of the aim of an artificial organ engineering book, we only report the main results.

Relations for mass transport coefficients are usually given in terms of dimensionless groups:

- Sherwood number, Sh, is defined as

$$Sh = \frac{k_c \ell}{\mathcal{D}_{AB}} \tag{2.7}$$

where ℓ is a characteristic length of the problem. Comparing the definition of Sherwood to the film model, the physical meaning of Sherwood number can be recognized as the ratio between the characteristic dimension, ℓ, and the film thickness, δ; usually, Sh is quite a large number, corresponding to a slender film layer;

- Reynolds number, Re, is defined as

$$Re = \frac{\rho v \ell}{\mu} \tag{2.8}$$

which represents the fluid motion conditions;

- Schmidt number, Sc, is defined as

$$Sc = \frac{\mu}{\rho \mathcal{D}_{AB}} \tag{2.9}$$

It is worth noting that Sc only contains fluid properties and represents the ratio between momentum diffusivity, μ/ρ, and diffusivity. For gases $Sc \approx 1$, while for liquids it is in the range $10^2 \div 10^3$.

It can be theoretically proved that mass transport coefficient can be expressed as a function of Re and Sc:

$$Sh = f(Re, Sc) \tag{2.10}$$

Such a function can be obtained by combining theoretical models (boundary layer, penetration theory, etc.) and correlations of experimental data. As an example, for a flow near a solid surface, with high velocity gradient near the surface, boundary layer model suggests a dependence of Sh on $Sc^{1/3}$; on the other hand, for mass transfer

Table 2.1 Correlation for mass transport coefficients

Fluid motion	Relations	Characteristic dimension, velocity
Inside circular pipe	$Sh = 0.023Re^{0.83}Sc^{1/3}$	Inner diameter of the pipe; mean velocity in the pipe
Parallel to flat plates	$Sh = 0.664Re^{1/2}Sc^{1/3}$	Plate length; approach velocity of the fluid
Past single solid sphere	$Sh = 2 + 0.6Re^{1/2}Sc^{1/3}$	Sphere diameter; fluid velocity far from the sphere
Past single gas sphere	$Sh = 2 + 0.645Re^{1/2}Sc^{1/2}$	Sphere diameter; fluid velocity far from the sphere
Through fixed beds of pellets	$Sh = 1.17Re^{0.58}Sc^{1/3}$	Pellet diameter; superficial velocity in the bed (volume flow rate divided by the empty bed section)

coefficient in a liquid phase near a gas phase (no velocity gradient in the liquid) penetration theory suggests a dependence of Sh on $Sc^{1/2}$.

Some correlations, referring to the more common flow configurations, are reported in Table 2.1.

2.4 Mass Transfer Between Two Phases

In the previous section, mass transport coefficients have been introduced to describe the mass transport occurring between an interface and the bulk of a homogeneous solution. In practice, we are more often interested in mass transfer between two contacting homogeneous phases; furthermore, in some cases a permeable membrane, offering a resistance to mass transfer, is placed between the two phases. As an example, let us consider oxygen transfer from air to blood, in direct contact or separated by a natural or artificial membrane.

In these cases, it may be desirable to directly describe the mass flux across the interface in terms of an appropriate difference in composition between the two phases and an overall mass transport coefficient. To this aim, we have to face two different problems:

- What is an appropriate difference in composition, to be used as the driving force for mass transfer?;
- How to relate the overall and single-phase mass transfer coefficients?

In the following, we discuss these two problems referring to mass transfer between a gas and a liquid phase in direct contact; we use partial pressure and concentration as composition variables for the gas phase and liquid phase, respectively; the reader can easily follow the same procedure to cope with different situations.

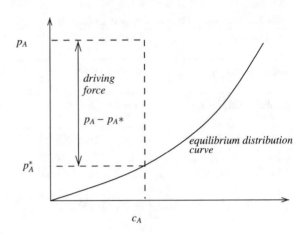

Fig. 2.2 Driving force for mass transfer in a gas–liquid system

As for the suitable composition difference, thermodynamics states that there is no mass transfer if the two phases are in equilibrium and, for a system not in equilibrium, mass transfer occurs to bring the system to equilibrium condition. Equilibrium conditions[2] are expressed in terms of an equilibrium distribution curve, i.e., a function that relates the composition in the gas phase (e.g., the partial pressure of A) with the composition of the liquid phase (e.g., the concentration of A) at a fixed temperature[3]

$$p_A = f(c_A) \tag{2.11}$$

Therefore, if we consider a two-phase system, with a gas at partial pressure p_A and a liquid with a concentration c_A (see Fig. 2.2):

- no transfer of A occurs if the partial pressure of A in the gas phase is equal to the partial pressure p_A^* corresponding to the equilibrium with c_A, i.e., if $p_A = p_A^* = f(c_A)$
- the driving force is the difference between p_A and $p_A^* = f(c_A)$; transfer occurs from gas to liquid if $p_A > p_A^*$ (gas absorption) or from liquid to gas if $p_A < p_A^*$ (stripping)[4]
- a suitable way to express the A flux through the interface may be

$$N_A = K_P(p_A - p_A^*) \tag{2.12}$$

---

[2]Thermodynamics states that equilibrium conditions are given by the equality of each component chemical potentials in the two phases.

[3]A side effect of this relationship is that we can refer to the composition of a phase α giving the composition of a different phase β in equilibrium with α: In medical field, it is usual to refer to oxygen concentration in the blood in terms of oxygen pressure, i.e., referring to the oxygen partial pressure in a gas phase in equilibrium with the blood.

[4]The driving force can be expressed also in terms of the concentration of A in the liquid phase: no transfer occurs if $c_{AL} = c_A^* = f^{-1}(p_A)$, where f^{-1} stands for the inverse of the function f, and the driving force is a difference between c_A and c_A^*; absorption occurs if $c_A < c_A^*$.

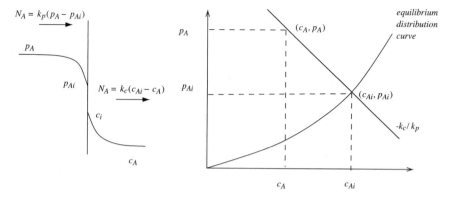

Fig. 2.3 Mass transfer from gas to liquid: flux and compositions at the interface

or

$$N_A = K_c \left(c_A - c_A^* \right) \tag{2.13}$$

K_P and K_c are two different overall mass transfer coefficients (with different units), but all of them sum up the resistances to mass transport in both phases. The next step is then related to the overall mass transport coefficient to the mass transport coefficients in gas and liquid phases.

To this aim, let us consider the absorption of A from a gas phase at p_A to a liquid phase at c_A $(p_A > p_A^*)$, as in the scheme reported in Fig. 2.3; since A moves from gas to liquid, we have a flux of A from the bulk of the gas phase toward the interface and from the interface to the bulk of the liquid phase; this implies that the partial pressure of A falls from p_A in the bulk of gas to p_{Ai} at the interface, where gas and liquid are in contact; similarly, in the liquid phase, the concentration of A drops from c_{Ai} at the interface to c_A in the bulk of liquid. If the diffusional resistances are only those in the fluid phase, the partial pressure and the concentration of A at the interface (p_{Ai} and c_{Ai}) are in equilibrium, i.e.,

$$p_{Ai} = f(c_{Ai}) \tag{2.14}$$

Furthermore, flux of A from bulk gas to the interface must equal flux of A from the interface to the bulk liquid

$$N_A = k_p (p_A - p_{Ai}) = k_c (c_{Ai} - c_A) \tag{2.15}$$

Equations 2.14 and 2.15 allow to get the interface composition and the flux of A, once the bulk composition and the single-phase mass transfer coefficients are known. The expression of the flux in terms of an overall mass transport coefficient is quite straightforward, if the equilibrium condition is a linear function, i.e., if at equilibrium, $dp/dc = m_{pc}$, with m_{pc} constant at least in the range of pressure and concentration we are interested in. In this case, it is possible to write

$$p_A - p_A^* = (p_A - p_{Ai}) + (p_{Ai} - p_A^*) = (p_A - p_{Ai}) + m_{pc}(c_{Ai} - c_A) \qquad (2.16)$$

and then, considering that $(p_A - p_{Ai}) = N_A/k_p$ and so on,

$$\frac{1}{K_p} = \frac{1}{k_p} + \frac{m_{pc}}{k_c} \qquad (2.17)$$

In a similar way, for K_c we get the following:

$$\frac{1}{K_c} = \frac{1}{m_{pc}k_p} + \frac{1}{k_c} \qquad (2.18)$$

The above equations show that the overall mass transfer coefficient depends on the single-phase mass transport coefficients and on the slope of the equilibrium distribution curve.

It may be interesting to consider the ratio

$$\frac{k_p}{K_P} = 1 + \frac{m_{pc}k_p}{k_c} \qquad \frac{k_c}{K_c} = 1 + \frac{k_c}{m_{pc}k_p} \qquad (2.19)$$

The term $m_{pc}k_p/k_c$ accounts for the relative importance of the mass transport resistances in each phase. More specifically:

- if $m_{pc}k_p/k_c \ll 1$ or $k_c/m_{pc}k_p \gg 1$, then $K_p \approx k_p$ while $K_c \approx m_{pc}k_p$ (see Fig. 2.4). In this case, the mass transport resistance in the liquid phase has a little effect on the overall mass transfer rate, i.e., mass transfer is *gas phase controlled*. Under such condition, any effort aimed at increasing the mass transfer rate should be directed to improving the gas-phase mass transport coefficient;

Fig. 2.4 Overall mass transfer coefficient as a function of the ratio $k_c/m_{pc}k_p$

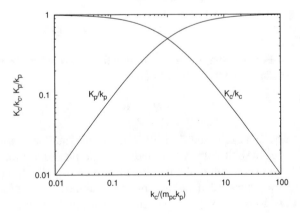

- if $m_{pc}k_p/k_c \gg 1$ $K_c \approx k_c$ and $K_p \approx k_c/m_{pc}$ (see Fig. 2.4). In this case, which occurs more frequently than the previous one, the mass transport resistance in the gas phase has a little effect on the overall mass transfer rate, i.e., mass transfer is *liquid phase controlled.* Under such condition, any effort aimed at increasing the mass transfer rate should be directed to improving the liquid-phase mass transport coefficient;

Symbols

c	Concentration
$\mathscr{D}_{AB}$	Diffusivity
J_A	Diffusive flux of the component A
k_c	Mass transfer coefficient, based upon concentration driving force
k_p	Mass transfer coefficient, based upon partial pressure driving force
k_x	Mass transfer coefficient, based upon mole fraction driving force
K_c	Overall mass transfer coefficient, based upon concentration driving force
K_p	Overall mass transfer coefficient, based upon pressure driving force
K_x	Overall mass transfer coefficient, based upon mole fraction driving force
ℓ	Characteristic length
m_{pc}	Slope of the equilibrium line, based upon partial pressure (gas phase) and concentration (liquid phase)
P	Pressure
p_A	Partial pressure of the component A
R	Gas constant
Re	Reynolds number
Sc	Schmidt number
Sh	Sherwood number
T	Temperature
v	Fluid velocity
x	Mole fraction
μ	Viscosity
ρ	Density

Subscripts

A	Related to component A
B	Related to component B
f	In the bulk of fluid
i	At the interphase

Superscripts

α	Phase α
β	Phase β
0	On the surface
*	Corresponding to equilibrium conditions

Chapter 3
Membrane Operations

3.1 Introduction

Membranes can be defined as thin layers (bidimensional structures) of insoluble materials that separate different phases or different regions within a liquid or gas phase. While impermeable membranes completely avoid any mass flow and act as a barrier between the two regions, semipermeable or permselective membranes allow to control mass transport between them. The main property of this latter type of membranes is their selectivity, i.e., the property to offer different resistance to the flow of different components, depending on their molecular size or chemical properties. As a result, components in a liquid or gas mixture can be separated on the basis of their different transport rate through a membrane.

In the human body, biological membranes perform numerous very important functions like separating the cell cytoplasm from the external environment (cellular membrane), or internal cellular organs from cytoplasm (nuclear membrane, mitochondrial membrane, etc.); the selectivity of such membranes is essential to maintain homeostasis and allow the flux of nutrients, ions, and gases. In the lungs, the alveolar membrane allows oxygen and carbon dioxide exchange between venous blood and air in the alveolar sacs, while avoiding direct contact between them. In the kidney, different membranes act to filtrate large amounts of plasma and low molecular weight solutes (glomerular filtration) from blood and, subsequently, reabsorb part of the filtered water, electrolytes, and other compounds via both passive and active transport processes (tubular transport).

Synthetic semipermeable membranes are used in separation processes in the biomedical field and, more generally, for industrial applications. The most important biomedical application of membranes is by far in hemodialysis devices, with a consumption of more than 200 million square meters of membranes by year [4]. In such process, blood flows in contact with a semipermeable membrane, while an aqueous rinsing solution flows on the other side of the membrane; urea, creatinine, and other small or medium molecular weight compounds are allowed to permeate across the membrane from the blood to the rinsing solution, while removal of larger molecules

© Springer-Verlag London 2017
M.C. Annesini et al., *Artificial Organ Engineering*,
DOI 10.1007/978-1-4471-6443-2_3

or particles (red and white cells, platelets, and blood proteins) is prevented. Gas permeable membranes are used in blood oxygenators to allow oxygen and carbon dioxide exchange between blood and air, without a direct contact between them. Finally, membrane-containing extracorporeal devices are used for blood detoxification (as in the artificial liver) or plasma separation.

While the application of membrane devices in the field of artificial organs is thoroughly covered in the second part of this book, this chapter presents the fundamentals of mass transport through membranes and a general description of different types of membranes available. The aim is to provide a theoretical framework for the analysis of the performance of separation units based on this technology.

This chapter is mainly focused on the use of membranes to separate specific compounds from liquid solutions; however, at the end of the chapter, membrane gas separation is also presented.

3.2   Phenomenological Aspects and Definitions

3.2.1   Membrane Types

Nearly, all the membranes used in biomedical devices are made of natural (cellulose derivatives) or synthetic (polysulfone, polymethyl methacrylate, etc.) polymers and can be broadly classified as hydrophilic or hydrophobic, depending on their tendency to be wet by water or reject water molecules.

Depending on their microstructural characteristics, membranes can be further classified into (see Fig. 3.1):

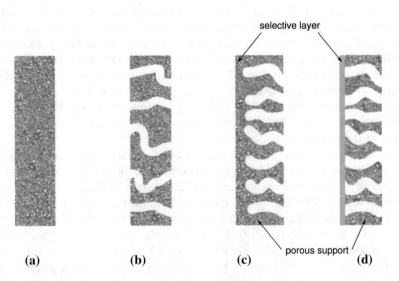

Fig. 3.1 Membrane types: **a** dense; **b** porous; **c** asymmetric; **d** composite

- Non-porous or dense membranes that are quite similar to very dense liquids and are homogeneous down to the near molecular scale; such membranes do not present pores of microscopic size on their surface.[1] All the components that can pass through the membrane have to dissolve in the dense material and move through it by Fick's diffusion (solution-diffusion mechanism). According to this transport mechanism, the selectivity of dense membranes is usually high, depending on the relative transport rate of the different components to be separated; on the other hand, mass flows are very low, unless the membrane is very thin. In order to ensure a sufficient rigidity and mechanical stability to withstand the pressure difference between the two sides of the membrane, a solid support is sometimes required.

- Porous membranes, whose structure is characterized by the presence of pores of different sizes (from several nanometers to several microns). In this case, the separation mechanism is based on sieving or filtration: solvent and small molecules can pass through pores, while larger molecules or suspended particles are retained. As for membranes with smaller pores, such as those used in ultrafiltration to separate high molecular weight compounds, the rating is usually given in terms of molecular weight cutoff (MWCO): a narrow pore size distribution results in a sharp MWCO with complete rejection of molecules larger than a certain size. While the selectivity of porous membranes is based only on the solute size, the fluxes through these membranes are usually high and large flow rates can be treated per unit membrane area.

- Asymmetric or composite membranes, with a very thin (typically from 0.1 to 1 μm) permselective layer supported on a highly porous layer ($\sim$30 μm): the non-porous selective skin determines the membrane selectivity, while the sponge-like support offers no resistance to mass transfer and has no effect on the membrane selectivity, but provides the mechanical strength required for membrane handling and to withstand pressure differences. Since the dense skin layer is very thin, the water flux through this type of membranes can be significantly higher than that obtained with dense homogeneous membranes. In asymmetric membranes, the same polymer forms both the permselective skin and the support layer, which only differ for porosity and pore size. In composite membranes, the porous support layer is coated with a different, highly cross-linked polymer that forms a very thin selective barrier.

3.2.2 Membrane Separation Processes

When a semipermeable membrane is used as a barrier to separate two aqueous solutions containing different compounds, water and/or solutes flow through the membrane as a consequence of a driving force (pressure, concentration, or electrical

---

[1]Pore size in dense membranes lies within the range of the thermal motion of polymer chains; widely accepted theories assume dense membranes to have no pores at all.

Table 3.1 Membrane separation processes

Driving force		Main flux and permeating compounds	Retentate	Process
Concentration gradient		Solvent + small molecules	Large molecules	Dialysis
Pressure gradient (bar)	0.5–2	Water + dissolved solutes, (proteins, amino acids, sugar, salts)	Microparticles: 0.1–10 m (cells, bacteria, colloids, emulsions)	Microfiltration
	1–10	Solvent + small molecules (salts, sugars, amino acids)	Macromolecules 5–50 nm (virus, proteins)	Ultrafiltration
	20–40	Solvent + small molecules (salts)	molecules 0.5–50 nm (proteins, vitamins, sugars)	Nanofiltration
	30–100	Solvent	Salts and low molecular weight solutes	Reverse Osmosis
Electrical potential gradient		Water + ionic solutes	Nonionic solutes	Electrodialysis

potential difference), depending on the physicochemical properties of the solute and characteristics of the membrane.

Membrane separation processes can be classified according to the type of driving force that determines the main mass flow and to the size range of molecules or particles separated by the membrane (see Table 3.1).

The term "dialysis" was coined in 1861 by Thomas Graham to describe the separation and selective filtration through semipermeable membranes. Today, this word indicates a process characterized by a solute flux driven by the difference of its concentration between the two sides of the membrane; according to the membrane properties, only some of the solutes present in the solutions can cross the membrane, while other solutes are retained; the solvent (water in aqueous solutions) flow through dialysis membranes is usually negligible. Dialysis is by far the most important membrane process used in biomedical devices: indeed, a dialysis process (conventional hemodialysis) is carried out to replace the kidney function in uremic patients.

Permeation occurs with a different mechanism when a membrane that is permeable to the solvent (water), but impermeable to the solutes separates two solutions at different concentrations. In this case, water crosses the membrane from the more diluted to the more concentrated solution (osmosis). Water flow can be reverted if a high enough transmembrane pressure gradient (higher than the osmotic pressure) is applied: as a result, water can be separated from the solutes. Depending on the size of the solute to be separated and membrane characteristics (pore size), the membrane processes driven by a pressure gradient are further classified as microfiltration, ultrafiltration, nanofiltration, or reverse osmosis (see Fig. 3.2).

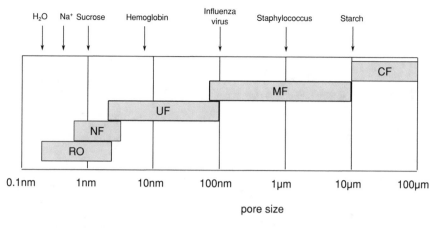

Fig. 3.2 Membrane processes for separation of solute of different size. RO: reverse osmosis; NF: nanofiltration; UF: ultrafiltration; MF: microfiltration; CF: conventional filtration. The size of some different solutes is also indicated

In the biomedical field, hemofiltration or hemodiafiltration is carried out in renal replacement therapy, to obtain a simultaneous flow of water and low molecular weight compounds from the blood compartment and to improve blood detoxification compared to pure hemodialysis. Finally, in electrodialysis, a gradient of electrical potential is used to separate charged compounds that migrate through the membrane driven by an electric force.

Hydrostatic pressure can also be used as driving force for gas separation, using both dense or microporous membranes.

3.2.3 Flow Patterns

It is worth noting that membrane separation process can be carried out in different modes:

- "dead end" mode, with the feed flowing perpendicularly to the membrane; this flow configuration can be used when the membrane is permeable to the solvent, while solutes or solid particles are retained and accumulate upstream of the membrane. Obviously, this flow configuration requires batch operation, with periodical removal of the retentate, also because the permeation rate decreases as the retentate concentration increases and particles accumulate on the upstream surface of the membrane.
- tangential operation with the feed flowing parallel to the membrane surface and the permeating species passing through the membrane. In this case, it may be necessary to feed a rinsing stream to the permeate side of the membrane, in cocurrent or countercurrent flow, to carry away the permeating compounds and enhance the

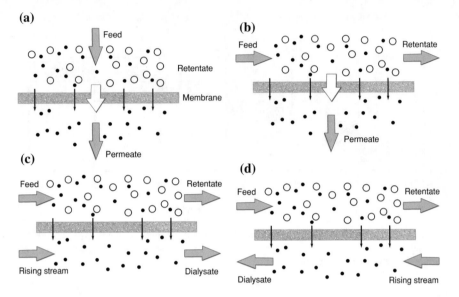

Fig. 3.3 Flow patterns in membrane devices. **a** Dead-end. **b** Cross-flow. **c** Co-current mode. **d** Counter-current mode

driving force for mass transfer. In cross-flow mode, the pressure gradient causes also a significant solvent flow through the membrane and the rinsing solution is not required (Fig. 3.3).

3.2.4   Membrane Modules

Membranes are assembled in different types of modules, which generally attempt to achieve a high membrane surface to module volume ratio. Among the different options available (plate-and-frame, spiral wound, tubular, hollow fiber), the most widely used membrane modules in biomedical devices are of the hollow fiber type. In hollow fiber membrane modules, thousands of small hollow fiber membranes, typically with a diameter of 300 μm and a wall thickness of 10 μm, are assembled inside a cylindrical shell (see Fig. 3.4); mass exchange occurs between two fluid phases that flow inside the fibers and in the shell space outside of the fibers, respectively. These modules offer a very high surface area per unit volume (a important feature, especially in biomedical devices), and the very small fiber diameter allows to sustain high transmembrane pressure gradients.

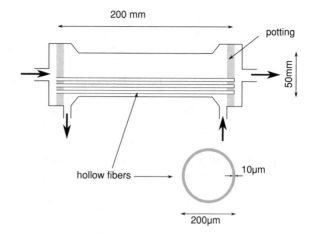

Fig. 3.4 Hollow fiber module. The sizes reported in the figure refer to a typical hollow fiber module used in hemodialysis

3.2.5 Osmosis, Osmotic Pressure, and Reverse Osmosis

As mentioned in Sect. 3.2.2, when a membrane permeable to solvent but impermeable to the solute separates a pure solvent from a solution, the solvent diffuses through the membrane toward the solute compartment. This phenomenon is called *osmosis* and has a considerable importance in many biological processes. As the solvent permeation proceeds, the pressure in the solution increases until an equilibrium condition is attained in which pressure is constant and no solvent flow is observed. This ideal experiment, schematically represented in Fig. 3.5, provides an operational definition of *osmotic pressure*, as the pressure which should be applied to the solution to prevent solvent permeation, when pure solvent is on the other side of the membrane.

A rigorous definition of the osmotic pressure is given in terms of chemical potential of the solvent in the solution [5–7]. In the above-described experiment, solvent flows from the higher (pure solvent) to the lower (solution) chemical potential side of the membrane; the flow stops when the chemical potentials of the solvent in the two

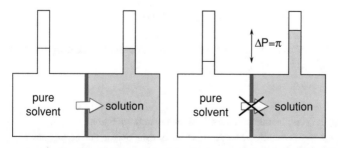

Fig. 3.5 Osmosis: If the pure solvent and the solution are at the same pressure, solvent flows toward the solution; to stop the solvent flow, the solution pressure must be higher than in the pure solvent solution. The pressure difference, π, is the osmotic pressure of the solution

compartments are equal, i.e., when the two phases are in thermodynamic equilibrium. Since chemical potential depends on composition, pressure, and temperature (which will be assumed as uniform), the equilibrium condition can be written as follows

$$\mu_w^0(T, P) = \mu_w(T, P + \pi, x_w) \tag{3.1}$$

where μ_w^0 is the chemical potential of the pure solvent (water), and x_w is the solvent molar fraction in the solution. Equation 3.1 can be assumed as the thermodynamic definition of osmotic pressure, π.

By the definition of the solvent *activity* in the solution, a_w [6], the right-hand side of Eq. 3.1 can be expressed as

$$\mu_w(T, P + \pi, x_w) = \mu_w^0(T, P) + v_w\pi + RT \ln a_w \tag{3.2}$$

where v_w is the molar volume of the solvent[2], and R the gas constant. By substituting Eq. 3.2 in Eq. 3.1, the following expression is obtained for the osmotic pressure:

$$\pi = -\frac{RT}{v_w^0} \ln a_w \tag{3.3}$$

For diluted ideal solutions,[3] Eq. 3.3 can be written as (van't Hoff law):

$$\pi = RT c_s \tag{3.4}$$

where c_s is the molar concentration of the solute. Equation 3.4 indicates that in diluted solutions, osmotic pressure does not depend on the characteristics of the solute; as a consequence, when several solutes are present, the osmotic pressure depends on the total concentration of all solutes, regardless of the specific composition of the solution. Furthermore, in the presence of dissociating solutes, as in the case of a strong electrolyte that completely dissociates in water, osmotic pressure is affected by the concentration of all species in the solution, including those deriving from the dissociation; in this case, it is convenient to write Eq. 3.4 as:

$$\pi = v RT c_s \tag{3.5}$$

where v is the number of species (ionic or molecular) originating from the dissociation of the solute.[4]

---

[2]Rigorously, the partial molar volume of the solvent in the solution.

[3]The water activity is usually expressed in terms of osmotic coefficient, Φ_w, as $\ln a_w = \Phi_w \ln x_w$; as for diluted solutions $\ln x_w = \ln(1 - x_s) \simeq -x_s = c_s/v_w$ so that $\pi = \Phi_w RT c_s$.

[4]For weak electrolytes, which are only partially dissociated in water with a degree of dissociation α, the concentration of species in solution is $c_s[1 + \alpha(v - 1)]$; therefore, the osmotic pressure of a dilute solution containing a weak electrolyte is $\pi = RT c_s[1 + \alpha(v - 1)]$.

Deviations from Eqs. 3.4 and 3.5 can be observed in concentrated solutions, when the ideal solution hypothesis is not acceptable. In these cases, the osmotic pressure can be expressed as a series expansion in c_s:

$$\pi = RT\, c_s (1 + B \cdot c_s + C \cdot c_s^2 + \cdots) \tag{3.6}$$

where $B, C \ldots$, are the second, third, etc., *osmotic virial coefficients*.

When two solutions are separated by a solvent-permeable membrane, solvent flows spontaneously across the membrane against its chemical potential gradient (i.e., from the higher to the lower chemical potential solution), so that permeation occurs with an overall decrease of the Gibbs free energy of the system, G. Indeed, if we consider the transfer of an infinitesimal amount of solvent moles dn_w from a solution 1 to a solution 2 at the same temperature, the overall infinitesimal Gibbs free energy change can be expressed as

$$dG = dn_w \left(\mu_{w,2} - \mu_{w,1}\right) = -v_w (\Delta P - \Delta\pi)\, dn_w \tag{3.7}$$

where $\Delta P = P_1 - P_2$ and $\Delta\pi = \pi_1 - \pi_2$ and subscripts 1 and 2 refer to the two solutions. Equation 3.7 shows that solvent transfer from solution 1 to solution 2 ($dn_w > 0$) is spontaneous ($dG < 0$) if $\mu_{w,1} > \mu_{w,2}$; furthermore, Eq. 3.7 shows that the sign of dG and, thus, direction of solvent flow depend on both the osmotic and hydrostatic pressure differences between the two solutions. More specifically, since the process occurs with a reduction of the free energy ($dG < 0$), the solvent flows from solution 1 to solution 2 ($dn_w > 0$) if $P_1 - P_2 > \pi_1 - \pi_2$ (or $P_1 - \pi_1 > P_2 - \pi_2$); however, solvent flow can be reversed ($dn_w < 0$) if $P_1 - P_2 < \pi_1 - \pi_2$ (or $P_2 - \pi_2 > P_1 - \pi_1$). This latter process is called *reverse osmosis*. Figure 3.6 shows a qualitative example of the hydrostatic and osmotic pressure levels in an osmosis ($P_1 = P_2$) and reverse osmosis process.

Even for diluted solutions, the osmotic pressure can be quite high. According to Eq. 3.4, at 25 °C $RT = 24.4$ l atm mole^{-1}, a 0.1 M solution has an osmotic pressure of 2.4 atm; therefore, high pressures are generally used in reverse osmosis processes (see Table 3.1).

Fig. 3.6 Osmosis and reverse osmosis processes

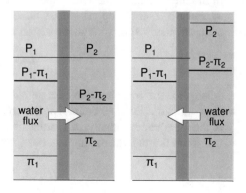

Finally, it is worth noting that the osmotic pressure is a measure of the water activity in the solution and, in diluted solutions, of solute concentration. Actually, the solute concentration can be measured as osmolarity (Osm), i.e., the molarity of an aqueous solution of an undissociated solute that provides the same osmotic pressure of the solution considered. So, for example, the osmolarity of a 1 M solution of glucose (which does not dissociate in water) is 1, while the osmolarity of a 1 M solution of NaCl (completely dissociated) is 2 Osm; healthy blood plasma is about 0.28 Osm, with an osmotic pressure of about 7.1 atm.

Referring to the osmotic pressure of cytoplasm, a solution can be defined as:

- Isotonic, if it has the same osmotic pressure of cell cytoplasm; an isotonic NaCl solution contains about 9 g/l of salt, with an osmotic pressure of about 6 atm;
- Hypertonic, if it has a greater osmotic pressure than cytoplasm; therefore, when an hypertonic solution is contacted with a cell membrane, water flows out of the cell causing the cell squeezing;
- Hypotonic, if it has a lower osmotic pressure than cytoplasm; if a cell is exposed to an hypotonic solution, water flows inward causing cell swelling.

In biomedical literature, it is customary to refer to the *oncotic pressure*, as the contribution to the osmotic pressure given by macromolecules (mainly proteins) or colloid particles present in the solution. In blood plasma, the high protein concentration (mainly albumin) results in an oncotic pressure of 25–30 mmHg that tends to draw water from the tissues into capillaries.

3.3 Mass Transport Through the Membrane

3.3.1 Flux Definitions

Transmembrane differences in solute concentrations and hydrostatic pressure cause different types of fluxes across the membrane:

- mass flux of each component, defined as the mass flow rate per unit surface area of the membrane; in particular, the solvent (water) flux, J_w, and the solute flux, J_s, must be considered. The mass fluxes are related to the velocity of different components flowing through the membrane as

$$J_i = c_i u_i \qquad (3.8)$$

where c_i is the concentration of component i and u_i its velocity.
- the volumetric flux, J_v, defined as the volumetric flow rate per unit surface area of the membrane; actually, J_v is equivalent to the bulk velocity, u, of the solution that flows through the membrane. For a diluted solution:

$$J_v \simeq J_w/c_w = u_w \qquad (3.9)$$

- the diffusive flux of the solute, J_i^D, defined as

$$J_i^D = c_i \, (u_i - u) \tag{3.10}$$

For a diluted solution,

$$J_s^D \simeq J_s - J_w c_s / c_w \tag{3.11}$$

It is worth noting that in a two-component system, only two independent fluxes can defined.

3.3.2 Fluxes, Driving Forces, and Transport Properties of the Membrane

Let us consider a semipermeable membrane placed between two compartments, containing two aqueous solutions at different concentrations. Each component (water or solute) flows through the membrane, in response to an overall driving force determined by the gradient of its chemical potential; actually, since the chemical potential depends on pressure and concentration, two driving forces, resulting from differences in pressure and concentration (or osmotic pressure), can be considered.

The general expressions to relate the fluxes to the driving forces are provided by thermodynamics of irreversible processes [8]; it is assumed that the membrane is not far from equilibrium so that a linear relationship holds between fluxes and driving forces. In terms of volumetric flux and solute flux, we get:

$$J_v = L_p (\Delta P - \sigma \, \Delta \pi) \tag{3.12}$$

$$J_s = \bar{c}_s (1 - \sigma) J_v + \omega \Delta \pi \tag{3.13}$$

where $\bar{c}_s$ is the mean solute concentration across the membrane; Eqs. 3.12 and 3.13 define three phenomenological coefficients, L_p, σ and ω that describe the transport properties of the membrane in the considered solvent/solute system. In particular:

- L_p is the *hydraulic permeability* of the membrane; if both the upstream and downstream compartments contain pure water, L_p represents the ratio between the volumetric flux through the membrane and the hydraulic pressure difference. L_p is equivalent to the ultrafiltration coefficient considered in hemodialysis (see Chap. 7); furthermore, the inverse of L_p may be viewed as a hydraulic resistance of the membrane.
- The *diffusive permeation coefficient* of the solute, ω, represents the ability of the solute to diffuse through the membrane; in the case of no volume flux ($J_v = 0$) ω represent the ratio between the solute flux and the difference in the osmotic pressure.
- The *reflection coefficient* (or Staverman coefficient), σ, which varies from 0 to 1, represents the coupling between the solvent and solute fluxes and measures how

easily the solute can bypass the membrane by convection: for an ideal semipermeable membrane, $\sigma = 1$ and the solute cannot cross the membrane if $\Delta\pi = 0$;
on the other hand, if the solute is unhindered, $\sigma = 0$ and the solute flows through
the membrane carried by the convective flux of the solvent even when $\Delta\pi = 0$. In
the biomedical literature, the sieving coefficient $S_\infty = 1 - \sigma$ is often considered
instead of σ.

Usually, Eq. 3.13 is rewritten in terms of solute concentrations, assuming that the
van't Hoff equation (Eq. 3.4) holds; therefore, in the last term in Eq. 3.13 $\omega\Delta\pi = \omega RT\Delta c_s$.

In order to deal with large concentration differences between the two phases
separated by the membrane, the above defined expressions for the solute flux can be
written in local form as

$$J_s = c_s(1 - \sigma)J_v - \mathscr{P}\frac{dc_s}{dz} \qquad (3.14)$$

where the membrane *permeability*, $\mathscr{P}$, was introduced.[5] Equation 3.14 shows in a
rather classical form the convection and diffusion components of solute flux: external
forces, like the pressure gradient, cause a bulk fluid motion with velocity J_v; the solute
is transported by convection according to the bulk velocity and diffusion according to
local concentration gradient. Integration of Eq. 3.14 along the membrane thickness
with the conditions[6] $z = 0$, $c_s = c_{s1}$ and $z = \delta$, $c_s = c_{s2}$ gives:

$$J_s = \frac{\mathscr{P}Pe}{\delta}\left(\frac{e^{Pe}}{e^{Pe} - 1}c_{s1} - \frac{1}{e^{Pe} - 1}c_{s2}\right) \qquad (3.15)$$

where the transmembrane Peclet number, which measures the convective to diffu-
sive transport rate ratio for the solute across the membrane, is defined as $Pe = (1 - \sigma)J_v\delta/\mathscr{P}$. Two limiting cases can be considered:

- In a dialysis process, with no solvent flux through the membrane, $Pe = 0$ and
  then

$$J_s = \frac{\mathscr{P}}{\delta}(c_1 - c_2) \qquad (3.16)$$

In this case, convection is absent and diffusion dominates solute transport across
the membrane;

- For high solvent flux ($Pe \to \infty$): $z = 0$, $c_s = c_{s1}$ and $z = \delta$, $c_s = c_{s2}$

$$J_s = J_v(1 - \sigma)c_1 \qquad (3.17)$$

In this case, the convective term dominates the solute transport.

---

[5] A comparison of Eqs. 3.13 and 3.14 gives $\mathscr{P} = \omega RT\delta$.

[6] Actually, the concentrations to be considered are those in the liquid phase in contact with the
membrane surface; such concentrations are generally different from those in the bulk of the fluid
phase, due to the resistance to mass transport in the boundary layers near the membrane surfaces
(see Sects. 3.5 and 3.5.2).

3.4 Physical Models for Membrane Transport Properties

Based on thermodynamics of irreversible processes, Sect. 3.3.2 presents a theoretical framework to describe mass transport through the membrane in terms of three coefficients that characterize the membrane behavior; however, no insight into the transport mechanisms in the membrane and dependence of transport parameters on the solution and membrane properties is provided; indeed, in the framework of thermodynamics of irreversible processes, the membrane is viewed as a black box and L_p, ω, and σ are phenomenological coefficients that, for each membrane-solute system, can only be obtained by experimental measurements.

Differently, physical models attempt to describe the mechanisms behind solvent and solute transport through the membrane and provide relationships between the transport parameters and the physicochemical properties of the membrane–solution system. There are numerous different membrane materials, and there are different ways in which the solution components can interact with the membrane; as a consequence, different physicochemical models have been proposed depending on the combination of membrane, solution, and operating conditions considered.

Two main different classes of physical models are used. The first one considers the membrane as a non-porous pseudo-homogeneous medium; different components dissolve in the membrane and diffuse against a concentration gradient (solution-diffusion model); differences in the component solubility and diffusivity in the membrane result in the membrane selectivity. The second type of models is the pore flow model, in which it is assumed that transport takes place through the membrane pores, both by diffusion and convection; in the simplest case, separation can be achieved because some solutes are excluded from some of the pores in the membrane. Solution-diffusion models and pore flow models are useful to describe mass transport in a membrane for both liquid and gas mixtures. In this section, we present the main aspects of the models that apply to the separation of liquid solutions; models for gas separation are discussed in Sect. 3.6.

3.4.1 Solution-Diffusion Model

In the solution-diffusion model, the membrane is viewed as a pseudo-homogeneous layer that can transmit pressure as a liquid; as a consequence, when a pressure difference is applied on the two sides of the membrane, the pressure is constant in the membrane at the high-pressure value, while it abruptly falls to the lower value, at the downstream side of the membrane. Transport of component i through the membrane is considered as a three-step process (see Fig. 3.7):

- component sorption in the upstream surface (side 1) of the membrane; thermodynamic equilibrium is achieved at membrane–fluid interface;
- molecular diffusion of the component in the membrane toward the downstream surface driven by the concentration gradient;

Fig. 3.7 Pressure and
concentration profiles across
the membrane in the
Solution-diffusion
permeation model

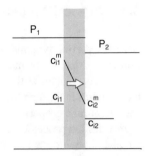

- desorption of the component at the downstream surface of the membrane (side 2); also in this case, thermodynamic equilibrium is assumed at the interface between the membrane and the permeate.

Since the pressure is uniform within the membrane, each component diffuses against its concentration gradient according to Fick's law:

$$J_i = -\mathscr{D}_i^m \frac{\mathrm{d}c_i^m}{\mathrm{d}z} \tag{3.18}$$

where $\mathscr{D}_i^m$ is the diffusivity of component i in the membrane. By integrating Eq. 3.18 along the membrane thickness, δ

$$J_i = \frac{\mathscr{D}_i^m}{\delta} \left(c_{i1}^m - c_{i2}^m \right) \tag{3.19}$$

where c_{i1}^m and c_{i2}^m are the solute concentrations in the membrane at the interface with the upstream and downstream fluid phases, respectively. These concentrations can be related to the concentrations in the fluid phases in contact with the membrane[7]

$$c_{i1}^m = k_i c_{i1} \tag{3.20}$$

and

$$c_{i2}^m = k_i c_{i2} \exp\left[\frac{-v_i \left(P_1 - P_2 \right)}{RT} \right] \tag{3.21}$$

where c_{i1} and c_{i2} are the concentrations of component i in the upstream and downstream liquid in contact with the membrane, respectively, k_i is the solubility coefficient of component i inside the membrane, and v_i is the partial molar volume of component i; it is worth noting that the exponential term in Eq. 3.21 derives from the pressure discontinuity at the permeate side of the membrane.

---

[7]As already underlined in the footnote 6, here, we refer to the concentrations and all the other properties of the liquid phases at the interface with the membrane; these properties may be different from those in the bulk fluid phases.

If the expressions for the concentrations in the membrane at the upstream and downstream faces are substituted into Eq. 3.19, the following expression for the flux across the membrane is obtained:

$$J_i = \frac{\mathscr{D}_i^m k_i c_{i1}}{\delta} \left\{ 1 - \frac{c_{i2}}{c_{i1}} \exp \left[\frac{-v_i (P_1 - P_2)}{RT} \right] \right\} \tag{3.22}$$

A significant simplification of Eq. 3.22 can be obtained in the relevant case of aqueous diluted solutions. In this case, two different expressions are obtained for water and diluted solutes.

As for water, accounting for the definition of the osmotic pressure,

$$\frac{c_{w2}}{c_{w1}} = \exp \left[\frac{v_w}{RT} (\pi_1 - \pi_2) \right] \tag{3.23}$$

and then

$$J_w = \frac{\mathscr{D}_w^m k_w c_{w1}}{\delta} \left\{ 1 - \exp \left[\frac{-v_w (\Delta P - \Delta \pi)}{RT} \right] \right\} \tag{3.24}$$

where $\Delta P = P_1 - P_2$ and $\Delta \pi = \pi_1 - \pi_2$. Equation 3.24 can be further simplified considering that

$$\exp \left[-v_w \frac{\Delta P - \Delta \pi}{RT} \right] \simeq 1 - v_w \frac{\Delta P - \Delta \pi}{RT}$$

so that

$$J_w = \frac{\mathscr{D}_w^m k_w}{\delta RT} (\Delta P - \Delta \pi) = A (\Delta P - \Delta \pi) \tag{3.25}$$

where A is usually defined as the *water permeability constant*.

As for the solute flow, since the term $v_i (\Delta P - \Delta \pi) / RT$ is small, it is possible to assume that the effect of the pressure difference is negligible and the driving force is almost equal to concentration difference. Therefore, the solute flux can be written as:

$$J_s = \frac{\mathscr{D}_s^m k_s}{\delta} (c_{s1} - c_{s2}) \tag{3.26}$$

It is worth noting that the model predicts a water flux that increases linearly with the applied pressure, while the solute flux is independent of pressure (see Fig. 3.8).

If we refer to a binary solution, a comparison of Eqs. 3.25 and 3.26 with the expressions derived from the thermodynamics of irreversible processes (Eqs. 3.12 and 3.13) shows that these are equivalent for an ideal membrane with $\sigma = 1$ with

$$L_P = \frac{\mathscr{D}_w^m k_w}{\delta RT} v_w \tag{3.27}$$

Fig. 3.8 Qualitative
dependence on pressure of
solvent and solute fluxes
through an ideal membrane
(solution-diffusion model)

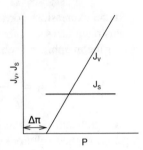

$$\omega = \frac{\mathcal{D}_s^m k_s}{\delta RT} \tag{3.28}$$

It is customary to define the membrane permeability for the solute as $\mathcal{P}_s = \mathcal{D}_s^m k_s$ and to write the solute flux:

$$J_s = \frac{\mathcal{P}_s}{\delta} (c_{s1} - c_{s2}) \tag{3.29}$$

The above equation shows that a high permeability results from both high solubility in the membrane (that usually depends on the chemical nature of the permeant and of the polymeric matrix) and a high diffusivity inside the membrane (that usually depends on the size of the permeating molecules).

3.4.2 Pore Flow Model

In porous membranes with well-defined pore structure, both solvent and solute flow occur through the pores that run through thickness of the membrane. Laminar flow of the solvent can be assumed within the membrane pores, due to their small size; therefore, for pure solvent flow, the velocity through the membrane can be given by the Hagen–Pouseuille law:

$$u_w = \frac{d_p^2}{32\mu} \frac{\Delta P}{\delta} \tag{3.30}$$

The general expression for the solvent velocity in the presence of solutes is obtained by introducing an effective pressure difference $\Delta P_e = \Delta P - \Delta\pi$ in Eq. 3.30; therefore, in this case, the water velocity is given by:

$$u_w = \frac{d_p^2}{32\mu} \frac{\Delta P - \Delta\pi}{\delta} \tag{3.31}$$

It is interesting to note that the viscous flow increases with the square of the pore diameter: if the membrane has a pore size distribution, the larger part of flow occurs through the larger pores.

Solute molecules larger than the pore size are rejected by the membrane. Solutes smaller than the pore size can penetrate within the pores, but size exclusion occurs: the solute molecules can only occupy the central part of the pore, at least a radius away from the pore wall, and as a consequence, the solute concentration in the solution inside the pores is lower than in the external environment. To account for size exclusion, a steric partition coefficient, K_s, is considered as a function of ratio between the molecule and the pore diameters, λ; for $\lambda < 1$,

$$K_s = (1 - \lambda)^2 \tag{3.32}$$

while if $\lambda \geq 1$, $K_s = 0$ [9]. Solutes that penetrate within the pores flow by convection, but correction coefficients must be included to account for the ratio between the solute and the pore diameters; therefore, the solute flux can be written as[8]:

$$J_s = K_v c_s^p u_w - \mathscr{D}_s K_D \frac{dc_s^p}{dz} \tag{3.33}$$

where c_s^p is the solute concentration in the pore and K_v and K_D are the hindrance coefficients that depend on λ [10]. For $\lambda \ll 1$ $K_v = K_D = 1$ and the classical convection diffusion equation is recovered; for larger molecules, $K_D < 1$, while $K_v > 1$ with a maximum of about 1.5 for $\lambda \sim 0.5$, since the large solute molecules tend to move near the center of the pore, where the velocity is higher.

The solute flux is then obtained by integrating Eq. 3.33 with the boundary conditions $z = 0$, $c_s^p = K_s c_{s1}$ and $z = \delta$, $c_s = K_s c_{s2}$.

3.5 Mass Transport in the Bulk Solution

In the previous sections, driving forces in flux expressions were given in terms of concentration or osmotic pressure differences between the two fluid phases at the interface with the membrane. Actually, the properties of the fluid phases at the membrane interface may be quite different from those of the bulk, due to the mass transport resistance in the fluid phases. In the design of membrane devices, the bulk properties of the fluid phases are known; therefore, it is interesting to relate the interphase concentrations and the fluxes to the concentrations in the bulk of the fluid phase.

To this aim, it is useful to analyze the mass transport in the fluid phase next the membrane with the film model. In such approach, the entire concentration difference

---

[8]If the effect of pressure on the chemical potential is considered, a term depending on the solute volume and pressure gradient is included in the diffusive term; this effect becomes more important as the effective pressure driving force increases [10].

is considered to be localized in a thin film of thickness δ, adjacent to the membrane surface; both convection and diffusion occur in the film, so that

$$J_s = c_s J_v - \mathscr{D}_s \frac{dc_s}{dz} \tag{3.34}$$

where $\mathscr{D}_s$ is the solute diffusivity in the fluid phase. At steady state, both J_s and J_v must be constant and equal to the fluxes through the membrane. With this condition, integration of Eq. 3.34 in the film gives

$$\ln \frac{J_s - J_v c_s^b}{J_s - J_v c_s'} = -\frac{J_v \delta_F}{\mathscr{D}_s} = -\frac{J_v}{k_c} \tag{3.35}$$

and

$$c_s' = c_s^b \left[1 - \frac{J_s - J_v c_s^b}{J_v} \left(e^{J_v/k_c} - 1 \right) \right] \tag{3.36}$$

where c_s^b and c_s' are the solute concentrations in the bulk of the fluid phase and at the interface with the membrane, respectively; the transport coefficient k_c is defined in this model as the ratio $\mathscr{D}_s/\delta_F$. Equation 3.36 shows that c_s' may be smaller or larger than c_s^b, depending on the value of $J_s - c_s^b J_v$. Let us consider the concentration profile in the upstream solution (phase 1). If the solute flux through the membrane is larger than the convective solute flux in the fluid film ($J_s > J_v c_{s1}^b$), then solute diffusive flow must be also directed toward the membrane; therefore, the solute concentration decreases in the film ($c_{s1}' < c_{s1}^b$). On the other hand, if the solute flux through the membrane is lower than the solute flux by convection in the fluid film ($J_s < J_v c_{s1}^b$), a backward diffusive flow in the film must occur to balance the flux in the film and in the membrane; therefore, a solute concentration gradient builds up in the film, with a concentration at the interface larger than that in the bulk of liquid ($c_{s1}' > c_{s1}^b$). A similar expression is derived for the downstream phase. The solvent flux is then expressed in terms of the bulk concentrations by substituting relations for the interphase concentrations in Eq. 3.15.

Different situations can occur depending on the flow scheme adopted, flow conditions, and physical properties of fluids. Two particular cases are worth being discussed in more detail and will be analyzed in the following sections: the first one is when there is no significant solvent flow through the membrane, as in conventional dialysis or in hemodialysis with tangential flow; the second one is when there is a large solvent flux, while the solute is at least partially rejected by the membrane.

3.5.1 Mass Transport in the Bulk Solution in Dialysis

Let us consider a dialysis process, with $J_v \simeq 0$ and J_s proportional to the difference between the solute concentrations at the membrane surfaces; the solute flux is then

Fig. 3.9 Driving forces in the membrane and in the liquid films at the interface with the membrane

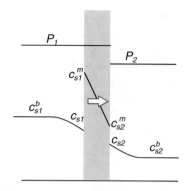

written as $J_s = \mathscr{P}\left(c'_{s1} - c'_{s2}\right)/\delta$, where, as already pointed out, the concentrations to be considered are those in the fluid in contact with the membrane surface. In this case, Eq. 3.36 simplifies as

$$c'_{s1} = c^b_{s1}\left(1 - \frac{J_s}{k_{c1}c^b_{s1}}\right) \tag{3.37}$$

or

$$J_s = k_{c1}\left(c^b_{s1} - c'_{s1}\right) \tag{3.38}$$

The resulting solute concentration profile is qualitatively reported in Fig. 3.9; the solute diffuses from the bulk of the upstream phase to the membrane surface and then passes through the membrane according to the solution-diffusion model and, eventually, diffuses from the downstream membrane surface to the bulk of the permeate.

It can be easily proved that the solute flux through the membrane can be expressed in terms of the mass transfer coefficients in the two fluid phases, k_{c1} and k_{c2}, and of the permeability of the membrane:

$$J_s = \frac{\left(c^b_{s1} - c^b_{s2}\right)}{\dfrac{1}{k_{c1}} + \dfrac{\delta}{\mathscr{P}} + \dfrac{1}{k_{c1}}} = K_c\left(c^b_{s1} - c^b_{s2}\right) \tag{3.39}$$

where c^b_{s1} and c^b_{s2} are the solute concentrations in the bulk of the upstream and downstream phases, respectively. From a different point of view, there are three resistances in series and the flux can be obtained in terms of an overall mass transfer coefficient, K_c, given by:

$$\frac{1}{K_c} = \frac{1}{k_{c1}} + \frac{\delta}{\mathscr{P}} + \frac{1}{k_{c2}} \tag{3.40}$$

It is worth noting that in this case, the solute flux through the membrane depends not only on the membrane properties ($\mathscr{P}$ and δ), but also on the mass transfer coefficients in the fluid phase, which in turn depend on the fluid dynamic conditions, as discussed in Chap. 2. In particular, the effect of the mass transfer coefficients in the fluid phase is significant if the solute permeability through the membrane is high. As for biomedical membrane devices such as hemodialyzers or membrane blood oxygenators, the improvements in membrane technology resulted in the production of highly permeable membranes, so that the blood film may become the controlling resistance for the mass transfer in these devices; efforts to improve the device performance are thus focused on fluid dynamic studies to improve the blood-side mass transfer coefficient.

3.5.2   Membrane Polarization

When solvent flow through the membrane is large and the solute is at least partially rejected, the solute accumulates near the upstream membrane surface, leading to a solute concentration increase in this region. This phenomenon, that is known as *membrane polarization*, is caused by the selectivity of the membrane itself. As discussed in the previous section, if the convective flux of the solute toward the membrane is larger than the solute flux through the membrane, a solute concentration gradient builds up in the feed-side liquid film and results in a backward solute diffusive flux; at steady state, the sum of the convective and diffusive solute fluxes must equal the solute flux through the membrane.

Let us focus the analysis on the limiting case in which the solute is completely rejected by the membrane ($J_s = 0$), so that Eq. 3.36 becomes:

$$c'_{s1} = c^b_{s1} e^{J_v/k_{c1}} > c^b_{s1} \tag{3.41}$$

Equation 3.41 shows that the larger the solvent flux, the larger the difference of solute concentration across the liquid film. The concentration profiles are qualitatively reported in Fig. 3.10.

The polarization effect is larger if the mass transfer coefficient is small as with high molecular weight solutes; therefore, membrane polarization is much more important in ultrafiltration rather than in reverse osmosis processes because of the low diffusivity of proteins and other high molecular weight compounds. The extent of concentration polarization can be reduced by enhancing the mass transfer coefficient near the membrane wall, e.g., by increasing the tangential velocity to promote turbulent flow.

As a consequence of the membrane polarization, increasing the transmembrane pressure difference results in an increase of the flux, smaller than that would be obtained with pure solvent; polarization increases the osmotic pressure of the solution next to the upstream face of the membrane, and referring again to a complete rejected solute, the solvent flux is given by

$$J_v = L_p \left(\Delta P - RT c'_{s1} \right) = L_p \left(\Delta P - RT c^b_{s1} e^{J_v/k_{c1}} \right) \tag{3.42}$$

Fig. 3.10 Membrane
polarization: qualitative
solute and solvent
concentration profiles in
permeate and retentate

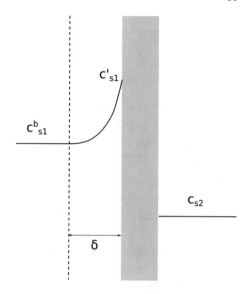

Actually, many experimental results show a further effect of membrane polarization: Increasing the pressure, the solvent flux levels off to a limiting value that seems to depend on the fluid dynamics of the liquid phase. To explain this, experimental evidence is usually assumed that the solute concentration next to the membrane becomes so high ($c'_{s1} = c_{gel}$) that the solute precipitates as a gel layer; the gel layer results in a further hydraulic resistance to the solvent flow, preventing it to further increase over the limiting value

$$J_{v,\max} = k_{c1} \ln \frac{c_{gel}}{c_{s1}} \tag{3.43}$$

As the pressure difference increases, the thickness (and the hydraulic resistance) of the gel increases maintaining the solvent flux to its limiting value:

$$J_{v,lim} = \frac{\Delta P - \Delta \pi}{\mathscr{R}_m + \mathscr{R}_{gel}} \tag{3.44}$$

where the hydraulic resistance of the membrane, $\mathscr{R}_m$, and of the gel, $\mathscr{R}_{gel}$, are considered.

3.6 Gas Permeation Through Membrane

Gas permeation occurs through both dense and porous membrane, and gas separation can result from differences in transport rate of different gases throughout the membrane. Two properties of the membrane are essential to determine the performance of a membrane device for gas mixtures: the membrane permeability that controls the throughput of the device or the gas flow rate per unit area of the membrane that can be achieved and the selectivity of the membrane that determines the ability to separate different gases or to achieve a fixed ratio between the fluxes of two different components.

As in liquids, gas transport through a membrane occurs with different mechanisms, according to the membrane structure: in dense, non-porous membrane, solution diffusion is widely accepted to be the main mechanism of transport, while in the porous membrane, gas transport is considered to occur via convection and diffusion.

3.6.1 Solution-Diffusion Model for Gas Permeation

The solution-diffusion model is widely used to describe the gas permeation through a dense, homogeneous membrane. With the same assumptions reported in Sect. 3.4.1 (constant pressure inside the membrane, equilibrium at the interface between the membrane and the fluid phases), the flux of a gas component through the membrane can be expressed by Eq. 3.19.

At the membrane surface, the gas concentration in the membrane can be related to the partial pressure in the gas phases by thermodynamic equilibrium conditions; if the Henry's law holds

$$c_i^m = k_i \, p_i \exp\left[-\frac{v_i \, (P - P_0)}{RT} \right] \simeq k_i \, p_i \tag{3.45}$$

where p_i is the partial pressure of component i in the gas phase in contact with the membrane, v_i the molar volume of i in the membrane phase and k_i is the solubility constant at pressure P_0; in the last term of Eq. 3.45, the effect of the total pressure in the equilibrium condition is neglected.

If the equilibrium conditions are substituted in the flux expression, we obtain:

$$J_i = \frac{\mathscr{D}_i^m k_i}{\delta} \, (p_{i1} - p_{i2}) = \frac{\mathscr{P}_i^G}{\delta} \left(p_i^F - p_i^P \right) \tag{3.46}$$

where the gas permeability coefficient, $\mathscr{P}_i^G = \mathscr{D}_i^m k_i$, is defined; it is customary to express $\mathscr{P}_i^G$ in Barrers[9] (1 Barrer $= 10^{-10} \, \mathrm{Ncm^3 \, cm^{-1} s^{-1} \, cmHg^{-1}}$). Beside permeability, the permeance, defined as the ratio $\mathscr{P}_i/\delta$, is also commonly used; permeance is usually expressed in gas permeability unity (GPU, 1GPU $=$ 1Barrer/cm).

As previously pointed out (see Sect. 3.4.1), the permeability coefficient (or the permeance) is the product of the diffusivity and solubility coefficient of the gas species; therefore, differences in permeability reflect differences in one or both

---

[9]It is worth noting the different dimensions of the permeability coefficients used for gases and liquids.

of these properties, which in turn depend on molecular size and physicochemical interactions between permeant and membrane. When the membrane is permeable to more than one component in the feed stream, the selectivity, defined as

$$\alpha_{AB} = \frac{\mathscr{P}_A^G}{\mathscr{P}_B^G} = \frac{\mathscr{D}_A^m}{\mathscr{D}_B^m}\frac{k_A}{k_B} \tag{3.47}$$

is an important parameter to be considered. Equation 3.47 shows that membrane selectivity between very similar gases (e.g., oxygen and nitrogen) can be due to the preferential diffusion of the smaller, higher-diffusivity gas; however, also when the molecular sizes and diffusivities are comparable, the membrane can be selective toward one component, if specific physicochemical interactions result in a higher solubility of this component in the membrane.

3.6.2 Gas Flow Through Porous Membranes

Gas flow through porous membranes can occur with different mechanisms. In the simplest case, where the gas flux is only due to the partial pressure gradient, but there is no difference in the total pressure between the two sides of the membrane, the flux of each component may be expressed by the Fick's model:

$$J_i = -\frac{\mathscr{D}_i}{RT}\frac{dp_i}{dz} \tag{3.48}$$

In porous membranes, two different types of diffusion, corresponding to different expressions for the diffusivity $\mathscr{D}$, can be considered:

- if the pores are not too narrow, gas diffusion in the pores occurs as in a free space and it is determined by the collisions between gas molecules; diffusivity inside the pore is the ordinary molecular diffusivity in free space. If a binary or pseudo-binary $i-j$ mixture is considered, according to the kinetic theory, the gas molecular diffusivity can be estimated as follows [1]:

$$\mathscr{D}_i = \mathscr{D}_{ij} = \sqrt{\frac{2}{\pi^3}}\frac{(RT)^{3/2}}{\mathscr{N}_A d_{c,ij}^2 P}\left(\frac{1}{M_i} + \frac{1}{M_j}\right)^{1/2} \qquad d_{c,ij} = \frac{d_{c,i} + d_{c,j}}{2} \tag{3.49}$$

where $\mathscr{N}_A$ is the Avogadro constant, M the molar mass, and d_c the collision diameter.
- at low pressure or in very small pores, the mean-free path of molecules is given by

$$\ell = \frac{3\mu}{2P}\left(\frac{\pi RT}{2M}\right)^{1/2} \tag{3.50}$$

where μ is the viscosity of the gas phase and can be comparable with the pore size; thus, the collisions of the molecules with the pore wall become more frequent than those between gas molecules. In such diffusion regime, referred to as *Knudsen diffusion*, diffusivity is given by:

$$\mathscr{D}_i = \mathscr{D}_{K,i} = d_p \sqrt{\frac{8}{9\pi} \frac{RT}{M_i}} \tag{3.51}$$

It is important to note that in this case, diffusivity is inversely proportional to the square root of the molecular weight of the gas.

In the transition region, where both types of diffusion play a significant role, $\mathscr{D}_i$ can be calculated as

$$\frac{1}{\mathscr{D}_i} = \frac{1}{\mathscr{D}_{ij}} + \frac{1}{\mathscr{D}_{K,i}} \tag{3.52}$$

If there is a pressure gradient, a viscous flow superimposes to diffusive transport in the pores and the flux can be expressed as:

$$J_i = -\frac{\mathcal{D}_i}{RT} \frac{dp_i}{dz} - \frac{B_0 p_i}{\mu RT} \frac{dp}{dz} \tag{3.53}$$

where B_0 is a geometric factor that accounts for the pore size and shape and is usually determined experimentally. For a cylindrical pore, $B_0 = d_p^2/32$.

Acronyms

CF	Conventional filtration
GPU	Gas permeation unit
MF	Microfiltration
MWCO	Molecular weight cutoff
NF	Nanofiltration
RO	Reverse osmosis
UF	Ultrafiltration

Symbols

a_i	Activity of the component i
A	Water permeability coefficient
B	Second osmotic virial coefficient in Eq. 3.6
B_0	Geometric factor for membrane pores
c_s	Solute concentration
c_s'	Solute concentration at the membrane interface in the fluid phase
d_c	Collision diameter
d_p	Pore diameter
C	Third osmotic virial coefficient in Eq. 3.6
$\mathscr{D}$	Molecular diffusivity

$\mathscr{D}_K$	Knudsen diffusivity
G	Gibbs free energy
J_i	Mass flux of the component i
J_v	Volumetric flux
J_i^D	Diffusive flux of the component i
K_v	Hindrance coefficient for solute convection in the pore
K_D	Hindrance coefficient for solute diffusion in the pore
K_s	Size exclusion partition coefficient
L_P	Hydraulic permeability
M	Molecular mass
$\mathscr{N}_A$	Avogadro constant
n_i	Number of moles of component i
P	Hydraulic pressure
p	Partial pressure
$\mathscr{P}$	Membrane permeability
Pe	Transmembrane Peclet number
R	Gas constant
T	Temperature
v_i	Molar volume of the component i
u	Convective velocity
u_i	Velocity of the component i
x_i	Mole fraction of the component i
δ	Membrane thickness
Φ_w	Solvent (water) osmotic coefficient
μ_i, μ_w	Chemical potential of the component i, chemical potential of (water)
μ	Viscosity
ν	Number of ions obtained from electrolyte dissociation
π	Osmotic pressure
σ	Reflection coefficient
ω	Diffusive permeation coefficient

Subscript

1, 2	Compartment 1, 2
i, j	Component i, j
s	Solute
w	Water (solvent)

Superscript

0	Pure component
F	Feed
m	Membrane phase
P	Permeate
p	Membrane pore

Chapter 4
Adsorption

4.1 Introduction

When a solution is contacted with another phase, solute concentration gradients may be observed in proximity of the interface, even at thermodynamic equilibrium. In particular, the concentrations of some solutes may be higher at the interface than in the bulk solution. This phenomenon is called *adsorption* and is caused by unbalanced interparticle forces at the interface of the two contacting phases.

Even if adsorption may occur at liquid–gas and liquid–liquid interfaces, as typically observed for surfactants, this term is much more commonly referred to the specific case of enrichment of one or more components in an interfacial layer at fluid–solid interfaces. Indeed, adsorption of solutes, also called *adsorbates*, on solid surfaces is the principle at the basis of many separation operations that are widely used both at the industrial and laboratory scale. Just to cite some examples, adsorption is used in bulk separation of mixtures of different classes of hydrocarbons, purification of natural gas from acid gases, removal of chlorine from tap water, and also in many analytical and preparative laboratory techniques based on chromatographic separation.

Adsorption has also been applied to blood purification and in the last decades played an important role in renal and liver replacement therapy. Adsorption has been historically used in the management of uremic patients with different approaches [11, 12], which include direct contact of plasma with solid sorbents to remove toxins (hemoperfusion), oral administration of solid sorbents to reduce the assimilation of nitrogen compounds, dialysate regeneration in peritoneal dialysis, and hemodialysis (see Chap. 7). The online regeneration by adsorption of the ultrafiltrate may be applied to avoid the use of exogenous replacement fluids in hemodiafiltration [13]. Furthermore, sorbents may be included in the dialyzer membranes in order to enhance the clearance of specific toxins [11].

Currently, adsorption appears as a promising route for removing compounds in the middle molecular weight range (8–15 kD), thus filling the gap between the scope of hemodialysis and hemofiltration treatments [12]. Furthermore, adsorption can

© Springer-Verlag London 2017
M.C. Annesini et al., *Artificial Organ Engineering*,
DOI 10.1007/978-1-4471-6443-2_4

nowadays be considered as the key separation process for the removal of water insoluble and strongly protein-bound plasma toxins, which are not effectively and selectively cleared by simple dialysis and/or ultrafiltration processes. For this reason, adsorption units are embedded in virtually all extracorporeal liver support devices used in the current clinical practice (see Chap. 8).

Adsorption and adsorption processes are a wide and complex subject thoroughly covered in many textbooks [14–17]. Far from being an exhaustive treatise, this chapter presents a short overview of the main aspects related to this topic. In consideration of the specific target of this book, only liquid purification operations, i.e., adsorption of one or more diluted species from a liquid solution, will be considered. Furthermore, since in most applications adsorption is carried out in fixed-bed columns, the chapter is focused on this operating mode. Particular emphasis is put on the use of mathematical models as a tool for gaining a deeper insight into the working principles and operation of adsorption units embedded in artificial organs. These models will be later used to discuss and compare different configurations and operating conditions of supportive devices used in organ replacement therapies.

4.2   Adsorptive Media

Several naturally occurring materials, such as clay, sand, and wood charcoal, have been used since antiquity for purifying liquid streams in applications such as desalination of water or dechlorination. Also the capability of some types of carbon to take up large amounts of gas has been long known. The first quantitative studies on these phenomena began at the end of eighteenth century, but it was only in the early twentieth century that a real insight was gained into the theoretical interpretation of the experimental observations. Since then, the industrial application of adsorption on the industrial scale started to develop, mainly for purification processes. During the second half of twentieth century, when adsorption began to be applied also to bulk separations, several new adsorbent materials have been developed together with novel laboratory techniques for their characterization.

Good solid adsorbents should in general exhibit a high *specific adsorption capacity* (i.e., quantity of adsorbate per unit mass of adsorbent at equilibrium), fast adsorption kinetics, and high selectivity for the components that must be adsorbed. The adsorption capacity depends on the thermodynamic affinity of the adsorbate for the solid and, since adsorption is a surface phenomenon, also on the specific surface of the solid. Therefore, in order to have a significant specific adsorption capacity, a solid must possess a highly porous structure. Three types of pores, corresponding to different scales of porosity, are conventionally defined on the basis of their size d_0: macropores ($d_0 < 50$), mesopores ($2\,\mathrm{nm} < d_0 < 50\,\mathrm{nm}$), and micropores ($d_0 < 2$). Adsorption kinetics are related to the solute transport rate within the solid particles. Diffusion of solutes through macropores is easier than in meso- and micropores, so that the presence of macroporosity is important to avoid excessively low adsorption rates. Selectivity of sorbents toward different adsorbates may be obtained through different mechanisms: by difference in thermodynamic affinity, adsorption kinetics, or because steric hindrance.

For most applications, a good adsorbent should also be easily regenerable. However, adsorption units used in blood detoxification are disposed after use for obvious safety reasons; therefore, this property is not required for this application.

Currently, the most commonly used adsorptive solids include activated alumina, activated carbons, silica gel, zeolites, polymeric resins, and carbon molecular sieves. The description of the nature, features, and applications of all these different classes of solids is beyond the scope of this book, and the reader is addressed to more specific textbooks for further information [14–17]. Rather, a short description will be provided here only for those adsorbents that are currently used in organ replacement therapies. These include

- **Activated carbons**. These widely used adsorptive solids are produced by high-temperature thermal decomposition (pyrolysis) of several carbonaceous materials, such as wood, coconut shells, and refinery residuals, just to cite some. Pyrolysis is coupled with an activation process, which involves controlled oxidation of the solid surface. Activation can be carried out either by physical processes, in which the pyrolyzed material is exposed to oxygen, carbon dioxide, or steam, or by impregnating the carbon with specific chemicals prior to pyrolysis. In this latter case, pyrolysis and activation are carried out simultaneously. Activated carbons are constituted by graphitic microcrystallites stacked with random orientation in an amorphous structure. The pore size distribution is bi- or trimodal, usually including both macro- and micropores. By tuning the production process, it is possible to obtain different pore size distributions and properties, thus covering a wide range of applications.

  Activated carbons have been used in hemoperfusion treatments for removing toxins by direct contact with blood. Due to their low biocompatibility and selectivity toward toxins, in such application activated carbons caused many side effects such as platelet depletion, hemolysis, hemorrhage, and hypotension [11]. Polymer-coated activated carbons have been developed in order to increase the biocompatibility. The polymer coating actually reduces the side effects but increases the adsorbate transfer resistance, thus slowing down adsorption kinetics. Uncoated activated carbons are still used in organ replacement therapies when no direct contact with blood is involved, like in the regeneration of the dialysate of kidney and liver support devices.

- **Nonionic polymeric resins**. Several nonionic polymers are used in industrial adsorption and analytical chromatography applications. The typical resins are based on copolymers of styrene/divinyl benzene and acrylic acid esters/divinyl benzene. Surface areas may be in excess of $750\,m^2/g$. Some usual applications of this type of adsorbents include the recovery of aromatics, chlorinated organics, proteins, and enzymes from aqueous solutions.

  Macroporous polystyrene/divinyl benzene resins are used in clinical practice for both hemoperfusion and regeneration of dialysate fluids.

- **Ion-exchange resins**. Several synthetic polymers contain functional groups that, when contacted with a liquid phase, can release ions by simultaneously taking up same-charge ions. These solids are commonly used in water-softening processes to remove Ca^{2+} and Mg^{2+}. Ion-exchange and adsorption are different phenomena, based on different physical mechanisms. However, ion-exchange and liquid adsorption processes present several similarities, so that the operation and control of adsorption and ion-exchange units is virtually identical.
  Ion-exchange resins are used in blood detoxification to remove charged toxins like bilirubin.

4.3 Adsorption Isotherms

In the classical framework of adsorption thermodynamics developed by Gibbs, adsorption is considered as the partition of components between a bulk fluid phase and a bidimensional surface phase (adsorbed phase). Like for any other phase equilibrium, one of the main problems of adsorption thermodynamics is to determine the composition of one phase once temperature (or pressure) and the composition of the other phase have been fixed. This information is of primary importance in the quantitative analysis and design of adsorption processes.

The composition of the adsorbed phase is usually expressed for each component as *specific adsorbed amount* n_i (amount of adsorbate per unit mass of adsorbent) or adsorbate concentration q_i (amount of adsorbate per unit volume of adsorbent). Clearly, $n_i = \rho_s q_i$, where ρ_s is the density of the solid adsorbent.

Adsorption equilibrium conditions are in general provided as a relation, at fixed temperature, between the specific adsorbed amount or concentration of one solute and the concentrations[1] of all the n_c components in the bulk fluid phase

$$q_i = f(C_1, \ldots, C_{n_c}) \quad [T] \tag{4.1}$$

A relation such as Eq. 4.1 is known as the *adsorption isotherm* of component i. The determination of an expression for f may be complex, especially if adsorption from concentrated liquid solutions, with non-negligible interactions between different components in both the liquid and adsorbed phase, is considered.

If the scenario is limited to the adsorption of a single solute from a dilute solution, the problem simplifies significantly. In this case, the adsorption isotherm reduces to a relation between the solute concentration in the bulk fluid and the specific adsorbed amount (or adsorbed concentration)

$$q_i = n_i/\rho_s = f(C_i,) \quad \text{or} \quad C_i = f^{-1}(q_i) \quad [T] \tag{4.2}$$

---

[1] For adsorption from gases, partial pressures are rather used.

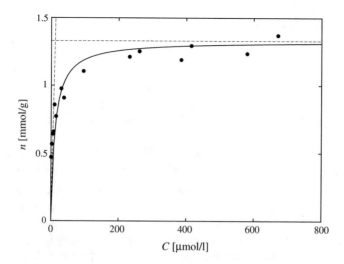

Fig. 4.1 Adsorption isotherm (25 °C) of tryptophan on activated carbon from diluted aqueous solution. *Points* experimental data from [18]; *Solid line* Langmuir isotherm fit ($n_{max} = 1.33$ mmol/g; $K_{lang} = 11.57$ μmol/l)

Figure 4.1 shows a set of constant temperature n_i versus C_i equilibrium data referred to the adsorption of a component from a diluted liquid solution. The shape observed in Fig. 4.1 is quite typical[2]: the specific adsorbed amount increases with the concentration in the liquid phase following a concave trend until a plateau is attained. Thus, the solid adsorbent exhibits a saturation limit for the adsorbate, corresponding to a *maximum adsorption capacity* n_{max} (or q_{max}).

In order to correlate adsorption data such as those reported in Fig. 4.1, several functional forms for f have been proposed, either on a purely empirical basis or in the framework of a physical model. One of the oldest and most commonly applicable expressions was proposed by Langmuir [20] and can be written as

$$q_i = q_{max,i} \frac{C_i}{K_{lang} + C_i} \tag{4.3}$$

where K_{lang} is a constant having the dimensions of a concentration. In the original derivation, Langmuir obtained Eq. 4.3 for the adsorption of a gas, by making the following assumptions:

---

[2]Though very common, the shape of the isotherm reported in Fig. 4.1 is not the only one that can observed experimentally. Lyklema [19] reports six possible classes of adsorption isotherms for dilute liquid solutions; however, the observed shape of the experimental adsorption isotherm depends also on the concentration range considered.

1. Adsorption occurs as a reversible chemical reaction between the gaseous component (A) and specific adsorption sites on the solid surface (S)

$$A + S \rightleftarrows AS \qquad (4.4)$$

2. Each site can accommodate one adsorbate molecule.
3. The solid surface is energetically uniform, i.e., adsorbate molecules have the same affinity for all sites.
4. Interactions between adsorbate molecules (lateral interactions) are absent; therefore, the heat of adsorption is independent of the solute adsorbed amount.

In this framework, $q_{max,i}$ is the concentration of adsorption sites on the solid and K_{lang} the reciprocal of the equilibrium constant of the adsorbate-site complex formation reaction (4.4). However, it should be noted that Eq. 4.3 can be obtained also by assuming different models of the adsorption mechanism, leading to a different physical interpretation of the parameters q_{max} and K_{lang}; furthermore, Eq. 4.1 provides a consistent representation of adsorption equilibrium data also for many systems that do not satisfy Langmuir's hypotheses. Indeed, Fig. 4.1 shows that the Langmuir isotherm provides a good fit of the data reported, even if they refer to adsorption from a liquid solution. In conclusion, it may be stated that the capability of an isotherm expression f to provide a good fit of experimental data may be due to its flexibility rather than physical soundness of the underlying theoretical framework. This applies equally to the Langmuir and other isotherm types.

It is worth considering two further adsorption isotherms that may be considered as limiting cases of the Langmuir isotherm:

- Linear isotherm

$$q_i = m_i C_i \qquad (4.5)$$

also known as Henry's law for adsorption. It is readily seen that the Langmuir isotherm (Eq. 4.3) approaches Eq. 4.5 at low solute concentrations

$$\lim_{C_i \to 0} \frac{q_i}{C_i} = \frac{q_{max}}{K_{lang}} = m_i \qquad (4.6)$$

The same behavior is exhibited by other (but not all) common adsorption isotherm expressions.

- Rectangular or highly favorable isotherm

$$q_i = \begin{cases} 0 & C_i = 0 \\ q_{max} & C_i > 0 \end{cases} \qquad (4.7)$$

In Langmuir's theoretical framework, this isotherm corresponds to an extremely favorable adsorption reaction, i.e., $K_{lang} \to 0$. Indeed, the Langmuir isotherm degenerates into Eq. 4.7 as K_{lang} approaches zero (Fig. 4.2).

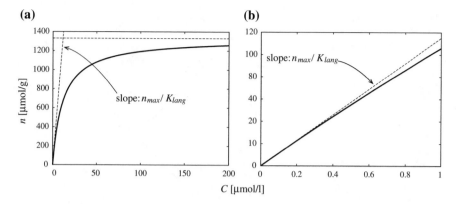

Fig. 4.2 Features of the Langmuir adsorption isotherm. *Dashed lines* represent the limiting behavior for $C \to 0$ and $C \to \infty$. On a wider concentration range (**a**), the concavity of the curve is important; if the same isotherm is considered only in the lower concentration range (**b**), the adsorption isotherm is nearly linear

Under a practical point of view, the operating concentration range should be accounted for when choosing an isotherm model for the design of an adsorption process. Figure 4.1 shows that adsorption data that are well described by a Langmuir isotherm on a wider concentration range may be equally well represented by a linear isotherm in the low concentration range.

By defining a reference solute concentration C_{ref} providing the order of magnitude of the process operating concentrations,[3] the reducibility of the Langmuir isotherm to the linear or rectangular form can be measured with the *linearity parameter* λ defined as follows

$$\lambda = \frac{K_{lang}}{C_{ref} + K_{lang}}$$

If $\lambda \simeq 1$, the linear isotherm can be used; on the other hand, if $\lambda \simeq 0$, the rectangular isotherm is applicable. Figure 4.3 reports the plot of the function

$$\frac{q}{f(C_{ref})} = \frac{C_i/C_{ref}}{\lambda + (1 - \lambda)C_i/C_{ref}}$$

and shows how the shape of the Langmuir isotherm changes with λ over the reference concentration range.

As noted earlier, when multiple solutes are adsorbed, obtaining a multicomponent adsorption isotherm such as Eq. 4.1 can be an involved problem. In the case of diluted solutions, the ideal adsorbed solution theory (IAS, [21]) provides a method for obtaining multicomponent adsorption isotherms once the single-solute isotherm

---

[3]The reference concentration could be the maximum solute concentration found in the considered process.

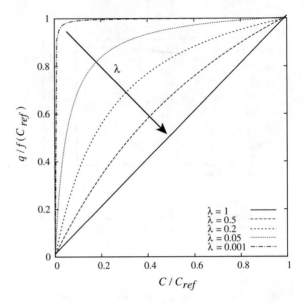

Fig. 4.3 Linearity parameter of a Langmuir isotherm: relation with isotherm shape

of each adsorbate is known. If all adsorbates follow a single-solute Langmuir isotherm (4.3) with the same $q_{\text{max},i} = q_{\text{max}}$ and different values of $K_{\text{lang},i}$, the application of the IAS theory results in the following expression

$$q_i = q_{\text{max}} \frac{C_i/K_{\text{lang},i}}{1 + \sum_{j=1}^{n_c-1} C_j/K_{\text{lang},j}} \tag{4.8}$$

where the solvent is the n_cth component in the mixture. Equation 4.8 can be obtained also by extending the adsorption mechanism hypothesized by Langmuir in deriving Eq. 4.3 to include the competition of multiple solutes for the same adsorption sites. By comparing Eqs. 4.3 and 4.8, it is readily seen that the adsorbed amount of a component is reduced in the presence of other adsorbates.

4.4   Mass Transfer Kinetics

When a fluid phase containing an adsorbable solute is contacted with an adsorbent solid, equilibrium conditions are not attained instantaneously. Since solute transfer occurs at a finite rate, the adsorbate concentration in the solid phase builds up progressively and not uniformly, until equilibrium is established and the composition gradients in the solid vanish. Mass transfer limitations are often important in adsorption processes and must be accounted for when designing an adsorption operation or choosing a sorptive material. It is indeed possible that a given material has a high capacity for an adsorbate, but exhibits very slow adsorption kinetics, so that the adsorbate uptake in the operation timescale is small. On the other hand, in some adsorption operation, different components in a mixture are separated selectively by exploiting the difference in their adsorption rates.

In order to account for finite mass transfer kinetics, adsorption operation models must include not only solute balances in the liquid and solid phases and equilibrium conditions, but also an expression of the solute flux transferred from the fluid to the solid phase. Such flux (referred to the external surface of the solid) will be hereafter denoted as N_{fs}.

This section presents two of the most common models used to calculate N_{fs} as a function of the instantaneous compositions of the two phases. The models presented assume that the solid is in the form of spherical particles of radius R_p. Furthermore, adsorption of a single solute will be hereafter considered, and the subscript i in concentrations and other component-specific properties will be omitted.

4.4.1 External and Intraparticle Mass Transfer Resistances

Transfer of the solute from the bulk fluid to the adsorbed phase involves three main steps:

- solute transfer from the bulk fluid to the solid particle surface across the surrounding fluid film;
- diffusion in the pores of the solid particle; and
- adsorption on the surface of the pores.

The latter step is often considered as instantaneous, so that the transfer rate is controlled by the first two steps. In this conditions, the transient evolution of the adsorbate concentration profile in a solid particle can be obtained by solving the microscopic mass balance

$$\frac{\partial q}{\partial t} = \mathscr{D}_p \frac{1}{r^2} \frac{\partial}{\partial r} \left(r^2 \frac{\partial q}{\partial r} \right) \tag{4.9}$$

where $\mathscr{D}_p$ is the solute effective diffusivity in the solid particles. Equation 4.9 must be integrated with an appropriate initial condition

$$q|_{t=0} = q_0(z) \tag{4.10}$$

and the boundary conditions

$$\left. \frac{\partial q}{\partial r} \right|_{r=0} = 0$$

$$k_c (C - C_s) = \mathscr{D}_q \left. \frac{\partial q}{\partial r} \right|_{r=R_p}$$

where C_s is the solute concentration in the liquid on the particle surface. By assuming equilibrium conditions at the interface,

$$q|_{r=R_p} = f(C_s) \tag{4.11}$$

The solute flux across the particle surface can therefore be expressed as

$$N_{fs} = k_c(C - C_s) = \mathscr{D}_p \left. \frac{\partial q}{\partial r} \right|_{r=R_p} \tag{4.12}$$

In most practical cases, the problem presented here requires a numerical solution, especially when Eq. 4.9 must be coupled with a solute balance in the liquid phase. It is worth considering two limiting cases in which important simplifications are obtained:

- **external mass transfer control**: Solute transport is much easier in the intraparticle space than in the external fluid film. In this case, radial concentration gradients in the solid particles are negligible and the adsorption rate is controlled by mass transfer resistance in the fluid film

$$C_S \simeq C \; ; \; N_{fs} = k_c(C - f^{-1}(q))$$

- **internal mass transfer control**: Solute transport is much easier in the external fluid film than in the intraparticle space. In this case, that is much more likely to be encountered in practice, the solute concentration in the fluid film is uniform and the adsorption rate is controlled by diffusion inside the solid pores

$$C_S \simeq C \; ; \; N_{fs} = \mathscr{D}_p \left. \frac{\partial q}{\partial r} \right|_{r=R_p}$$

Whether one of the above-mentioned simplifications is applicable depends not only on k_c (fluid dynamic and physical properties of the fluid phase) and $\mathscr{D}_p$, but also on the equilibrium conditions and operating concentration range. A dimensional analysis of the boundary condition (4.12) shows that the dimensionless particle Biot number

$$Bi_p = \frac{C_{ref}}{f(C_{ref})} \frac{k_c R_p}{\mathscr{D}_p}$$

can be used to detect a negligible mass transfer resistance. More specifically, if $Bi_p \ll 1$ or $Bi_p \gg 1$, the system is in the external or internal mass transfer control regime, respectively.

4.4.2   Linear Driving Force Model

Since its introduction [22], the *linear driving force* (LDF) model has been used frequently for the modelization of adsorption processes, mainly because of its formal and conceptual simplicity combined to physical consistency and sufficient flexibility.

The LDF model assumes that the solute flux transferred from the fluid to the solid phase can be expressed as

$$N_{fs} = k_{LDF}(q^* - \bar{q}) \tag{4.13}$$

where k_{LDF} is the *LDF mass transfer coefficient*, q^* is the solid phase adsorbate concentration in equilibrium with the fluid phase solute concentration, i.e., $q^* = f(C)$, and $\bar{q}$ is the volume-averaged adsorbate concentration in the solid particles

$$\bar{q} = \frac{1}{V_p} \int_{V_p} q \, dV = \frac{3}{R_p^2} \int_0^{R_p} q \, r^2 \, dr \tag{4.14}$$

The LDF mass transfer coefficient k_{LDF} is a lumped parameter that, from a general point of view, accounts for both internal and external mass transfer resistances. Therefore, this parameter is, in principle, related to k_c and $\mathscr{D}_p$. Nevertheless, because of the intrinsic semiempirical nature of the LDF model, this relation cannot be easily defined and k_{LDF} must be regarded as an adjustable parameter that can be obtained by fitting experimental data.

However, is it worth noting that in at least two cases, it is possible to define a relation between k_{LDF} and the physical properties of the system:

- If solute transfer is controlled by intraparticle diffusion ($Bi_p \gg 1$),

$$k_{LDF} \simeq 5\frac{\mathscr{D}_p}{R_p} \tag{4.15}$$

This result can be obtained by searching an approximate solution to Eq. 4.9 having the form[4]

$$q(r, t) = A(t) + B(t) \, r^2$$

with a finite volume approach [24].

---

[4]A similar procedure can be applied also by considering the wider class of approximate solutions

$$q(r, t) = A(t) + B(t) \, F(r) \tag{4.16}$$

where $F(r)$ is an arbitrary monotonous function that satisfies the condition

$$\left. \frac{\partial F}{\partial r} \right|_{r=0} = 0$$

In this case, a different expression of k_{LDF} is obtained [23].

- If solute transfer is controlled by the external film resistance ($Bi_p \ll 1$) and the equilibrium isotherm is linear,

$$k_{LDF} = \frac{k_c}{m}$$

where k_c is the solute mass transfer coefficient in the external fluid film.

4.5 Fixed-Bed Adsorption

In virtually all applications, both at industrial and laboratory scale, adsorption is carried out in the fixed-bed operating mode. In this configuration, the sorbent solid (stationary phase), usually in the form of porous beads, is packed inside a cylindrical column and constitutes the column fixed bed. The fluid solution (mobile phase) containing the species to be removed by adsorption is fed at one column extremity and crosses the column moving through the interstitial space of the column bed. During the contact with the solid phase, the adsorbable solutes are transferred to the solid phase so that, at the column outlet, the concentration of such species is lowered or zero. Clearly, the concentration of the adsorbates in the solid phase will increase until the solid loaded in the column attains equilibrium conditions with the fluid feed. At this point, no further removal of adsorbates from the fluid is possible, and the operation is no longer effective for the removal of the solutes.

In more detail, one typical operating cycle of an adsorption column is represented schematically in Fig. 4.4a

and can described as follows (a column removing a single adsorbate from a fluid stream is considered):

- The solid phase is initially free from adsorbates. Adsorption occurs in the first layers of the column bed close to the fluid inlet, where the adsorbate concentration in the fluid drops from the feed value to zero in the so-called *mass transfer zone* (MTZ). The mobile phase is free from solutes downstream of the mass transfer zone up to the column outlet.
- As the first layers of the solid bed reach equilibrium with the fluid feed, no further adsorption occurs and the solute concentration in the mobile phase remains constant in this part of the column. Therefore, the mass transfer zone moves progressively through the column and separates a part of the solid bed already in equilibrium with the column feed, from a part in which the solid bed is still clean.
- Eventually, the mass transfer zone will reach the column outlet. At this point, the whole solid bed will be in equilibrium with the fluid feed and the solute inlet and outlet concentration will coincide.

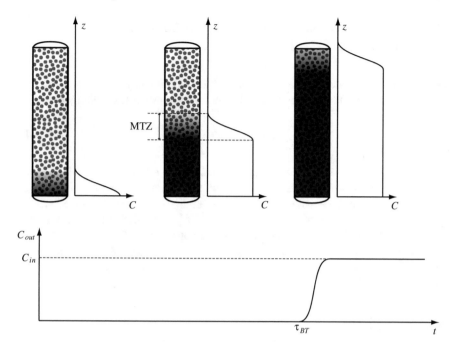

Fig. 4.4 Schematic description of the typical operation cycle of a fixed-bed adsorption column. *Top* advancement of the mass transfer zone (MTZ) through the column. *Bottom* breakthrough curve

At the end of the cycle, depending on the specific application, the sorbent solid is either regenerated (by removing adsorbates) and reused or disposed and substituted (possibly along with the whole column). For applications requiring continuous operation, multiple columns are operated in parallel and switched between adsorption and regeneration/solid substitution stages.

The solute concentration in the column effluent during one operating cycle is reported in Fig. 4.4b. This plot is called *breakthrough curve* and, as will be shown later, provides important information on the column behavior. Figure 4.4b shows that the solute is cleared from the fluid stream in the early operating time of the column; then, as the end of the cycle is approached and the solid bed is almost completely saturated with the adsorbate, the outlet solute concentration increases to the inlet value with a characteristic sigmoidal shape.

The adsorption cycle is stopped at *breakthrough time*, i.e., when the solute outlet concentration reaches a threshold value that is set according to the specific application requirements. Maximizing the breakthrough time while keeping the column size and solid mass to sensible values is clearly one of the main aims of the design of a fixed-bed adsorption operation, especially when disposable columns are used, as in biomedical supportive devices.

4.5.1 Model of a Fixed-Bed Adsorption Column with Liquid Feed

A mathematical model of a fixed-bed adsorption column may be built starting from the differential solute balance extended to an infinitesimal column element

$$\varepsilon \frac{\partial C}{\partial t} + (1 - \varepsilon) \frac{\partial q}{\partial t} = -\frac{\partial \langle N_z \rangle}{\partial z} \tag{4.17}$$

where t is time, z the axial coordinate in the column, ε the column bed void fraction, and $\langle N_z \rangle$ the surface-averaged axial solute flux on the column cross section. Equation 4.17 may be considered of general validity but cannot be solved without introducing further constraints among C, q, and $\langle N_z \rangle$, since all these variables depend on t and z. Different column models are obtained by coupling Eq. 4.17 with different expressions for $\langle N_z \rangle$ and kinetics of mass transfer between the solid and fluid phases. In the following part of this section, some of the most commonly used column models will be analyzed.

4.5.1.1 Ideal Column

An adsorption column characterized by

1. **plug flow of the mobile phase**, i.e., purely convective solute axial flux

$$\langle N_z \rangle = -uC \tag{4.18}$$

   where $u = Q/S$ is the liquid superficial velocity, and Q and S being the volumetric flow rate of the mobile phase and the column cross-sectional area, respectively.
2. **no mass transfer resistance between the liquid and solid phases**, i.e., instantaneous adsorption kinetics. Therefore, local equilibrium conditions hold throughout the column

$$q = f(C)$$

will be hereafter referred to as *ideal column*. For an ideal column, Eq. 4.17 becomes

$$\left[1 + \frac{1 - \varepsilon}{\varepsilon} f'(C) \right] \frac{\partial C}{\partial t} = -v \frac{\partial C}{\partial z} \tag{4.19}$$

where $v = u/\varepsilon$ is the interstitial velocity of the mobile phase. Equation 4.19 is a first-order quasi-linear homogeneous PDE that can be integrated with the two following boundary/initial conditions, corresponding to an initially clean column

$$C(z = 0) = C_{in}\ t \geq 0\ ;\quad C(t = 0) = 0\ \ 0 < z \leq L \tag{4.20}$$

where L is the length of the column bed.

First-order PDE theory is beyond the scope of this book, and the reader is addressed to specific texts for further details.[5] Here, only the results that are useful for understanding the column behavior are reported. The solution of Eq. 4.19 depends on the expression of the derivative $f'(C)$, i.e., on the equilibrium isotherm that applies to the system considered. Regardless of the specific expression, if the adsorption isotherm is concave[6] ($f''(C) \leq 0$, $C \geq 0$), the solution of Eq. 4.19 with conditions given in Eq. 4.20 takes the form

$$C(z, t) = \begin{cases} C_{in} & z \leq \zeta(t) \\ 0 & z > \zeta(t) \end{cases} \quad ; \quad \zeta(t) = v_f t$$

which means that in an ideal column with concave adsorption isotherm, the mass transfer zone has zero thickness and degenerates into a sharp front, whose axial position is denoted as ζ, which moves through the column with constant velocity $v_f = \zeta'(t)$.

When the saturation front reaches to the column outlet, the solute concentration in the column effluent, $C_{out} = C(z = L)$, changes abruptly from 0 to C_{in}. Therefore, the breakthrough curve in this case is a step function

$$C_{out}(t) = \begin{cases} 0 & t < \tau_{BT}^* \\ C_{in} & t \geq \tau_{BT}^* \end{cases}$$

and the ideal column breakthrough time, τ_{BT}^*, is clearly defined without the need to set a threshold concentration. The value of τ_{BT}^* can be determined by an overall solute balance from feed start until breakthrough time[7]

$$\tau_{BT}^* = \tau_b \left[1 + \frac{1 - \varepsilon}{\varepsilon} \frac{f(c_{in})}{c_{in}} \right] = \tau_b(1 + \mu) \tag{4.21}$$

where $\tau_b = L/v$ is the mobile phase residence time, and μ is the *capacity ratio*, that, for a given feed, is defined as the ratio between the maximum amount of solute that can be stored in the solid and liquid phases inside the column. Since the saturation front velocity is constant, Eq. 4.21 also implies that

$$v_f = \frac{v}{1 + \mu} \tag{4.22}$$

---

[5] A thorough treatise with specific applications to fixed-bed adsorption columns is proposed in [25].

[6] The linear and Langmuir isotherms satisfy this condition. The following results are applicable also for the rectangular isotherm, even if this isotherm is not differentiable for $C = 0$.

[7] That is by setting the amount of solute introduced in the column with the feed until breakthrough time to be equal to the overall amount of solute present in the column at that time

$$vSC_{in}\tau_{BT}^* = [\varepsilon C_{in} + (1 - \varepsilon)f(C_{in})]SL.$$

Equations 4.21 and 4.22 show that the saturation front moves slower than the mobile phase and the solute reaches the exit after the first fluid element has crossed the column. It could be told that the column "slows down" the solute compared to the mobile phase so that the column effluent is clean for a time τ_{BT}^*. The higher the adsorption capacity of the solid sorbent loaded in the column, the higher μ and the higher the breakthrough time. It is clear that a proper choice of the adsorbing material should in general imply $\mu \gg 1$.

4.5.1.2 Complete Column Model

The ideal column model provides sufficient insight into the basic features of an adsorption column behavior. However, such a model does not account for some important phenomena, whose effects are often non-negligible, that affect the shape breakthrough curve and may cause an important reduction of breakthrough time. More specifically, a complete column model should also account for:

1. **Axial dispersion**. On a microscopic scale, the fluid flow pattern in the column is quite complex. Fluid must go around solid particles following tortuous paths of different lengths; furthermore, the fluid velocity is not uniform in the gap between adjacent particles, because of the drag force exerted by solid boundaries. Despite such complexity, simple mass conservation considerations show that the average fluid velocity on the column section must be purely axial and constant throughout the column.

   The plug flow model only accounts for the average velocity field and assumes that this information is sufficient also for calculating the average solute convective flux $\langle N_z \rangle$ (Eq. 4.18). However, differences between local and average fluid velocities are not averaged-out when calculating $\langle N_z \rangle$; their effect, which is called *hydrodynamic dispersion*, is qualitatively (but not quantitatively) similar to those caused by molecular diffusion and is usually accounted for by letting

$$\langle N_z \rangle = -uC + \mathscr{D}_{dz} \frac{\partial C}{\partial z} \tag{4.23}$$

   where $\mathscr{D}_{dz}$ is the solute *axial dispersion coefficient*. Solute axial transport by molecular diffusion is usually negligible compared to convection, especially in liquids. However, the effect of molecular diffusion can be lumped into the dispersion term. Therefore, $\mathscr{D}_{dz}$ may depend also on (but does not coincide with) the molecular diffusivity of the solute in the fluid phase $\mathscr{D}_{mf}$. A review of methods for evaluating $\mathscr{D}_{dz}$ is presented in [16]. In general, correlations of the form

$$Pe_p = \frac{2R_p u}{\mathscr{D}_{dz}} = g(Re_p, Sc) \tag{4.24}$$

   are proposed. The particle Reynolds number Re_p and the Schmidt number Sc are defined as follows

$$Re_p = \frac{\rho_f 2R_p u}{\mu_f} \quad ; \quad Sc = \frac{\mu_f}{\rho_f \mathscr{D}_{mf}}$$

where ρ_f and μ_f are the density and viscosity of the fluid phase, respectively. In liquid systems, Pe_p has the order of magnitude of 1 and attains a maximum theoretical value of 2 at high Re_p.

By substituting Eq. 4.23 into Eq. 4.17, the solute balance in the column becomes

$$\varepsilon \frac{\partial C}{\partial t} + (1 - \varepsilon) \frac{\partial \bar{q}}{\partial t} = -u \frac{\partial C}{\partial z} + \mathscr{D}_{dz} \frac{\partial^2 C}{\partial z^2} \tag{4.25}$$

2. **Finite mass transfer kinetics**. In order to account for finite mass transfer kinetics, the column model must include an adsorbate balance in the solid phase. Such balance can be written for a single sorbent particle as

$$\frac{\partial \bar{q}}{\partial t} = a_p N_{fs} \tag{4.26}$$

where a_p is the specific interfacial area of a single solid particle. Therefore, for spherical particles of radius R_p, $a_p = 3/R_p$.

The complete column model is therefore composed of the two[8] differential equations reported in Eqs. 4.25 and 4.26. In this case, the conditions corresponding to an initially clean column are

$$q(t = 0) = 0 \ \ 0 \le z \le L \ ; \ \ C(t = 0) = 0 \ \ 0 < z \le L \tag{4.27}$$

and the boundary conditions for the fluid phase

$$uC_{in} = u \left. C \right|_{z=0} - \mathscr{D}_{dz} \left. \frac{\partial C}{\partial z} \right|_{z=0} \tag{4.28}$$

$$\left. \frac{\partial C}{\partial z} \right|_{z=L} = 0 \tag{4.29}$$

It is worth noting that the Danckwerts [26] boundary conditions (Eqs. 4.28 and 4.29), though widely used in this and similar types of problems, are arbitrary under certain points of view. Alternatives to such conditions are possible [27], but will not be considered here.

---

[8]A third differential equation, namely Eq. 4.9, must be included if the mass transfer model used accounts for intraparticle diffusion resistance.

4.5.1.3   Column with Axially Dispersed Flow, LDF Mass Transfer Kinetics, And Langmuir Adsorption Isotherm

The general model presented in Sect. 4.5.1.2 must be completed with the description of adsorption equilibrium and kinetics. If a Langmuir isotherm and LDF mass transfer kinetics are considered, Eqs. 4.3 and 4.13 will be included in the model.

In order to reduce the number of model parameters, the resulting system of differential equations can be conveniently cast in the following dimensionless form

$$\frac{1}{1+\mu}\frac{\partial \tilde{C}}{\partial \tilde{t}} + \frac{\mu}{1+\mu}\frac{\partial \tilde{q}}{\partial \tilde{t}} = -\frac{\partial \tilde{C}}{\partial \tilde{z}} + \frac{1}{Pe}\frac{\partial^2 \tilde{C}}{\partial \tilde{z}^2} \tag{4.30}$$

$$\frac{\partial \tilde{q}}{\partial \tilde{t}} = (1+\mu)St\left(\frac{\tilde{C}}{\lambda + (1-\lambda)\tilde{C}} - \tilde{q}\right) \tag{4.31}$$

$$\tilde{q}(\tilde{t}=0) = 0 \ \ 0 \le \tilde{z} \le 1 \ ; \ \ \tilde{C}(\tilde{t}=0) = 0 \ \ 0 < \tilde{z} \le 1 \tag{4.32}$$

$$1 = \hat{C}\Big|_{\tilde{z}=0} - \frac{1}{Pe}\frac{\partial \tilde{C}}{\partial \tilde{z}}\Big|_{\tilde{z}=0} \tag{4.33}$$

$$\frac{\partial \tilde{C}}{\partial \tilde{z}}\Big|_{\tilde{z}=1} = 0 \tag{4.34}$$

where the following dimensionless variables and parameters have been introduced

$$\tilde{z} = \frac{z}{H} \ \ \tilde{t} = \frac{t}{\tau_{BT}^*} \ \ \tilde{C} = \frac{C}{C_{in}} \ \ \tilde{q} = \frac{\bar{q}}{f(C_{in})} \ \ Pe = \frac{uL}{\mathcal{D}_{dz}} \ \ St = \tau_b a_p k_{LDF} \ \ \lambda = \frac{K_{lang}}{K_{lang} + C_{in}}$$

The modified Peclet number Pe gives a measure of the ratio between the rates of axial solute transport by convection and dispersion; therefore, if $Pe \gg 1$, axial dispersion can be neglected and a plug flow condition is approached in the column. The modified Stanton number St gives a measure of the ratio between the rates of adsorption and axial convection in the column.

4.5.2   Analysis of Breakthrough Curves

In this section, the model presented in Sect. 4.5.1.3 is applied to the calculation of breakthrough curves of an adsorption column, with the aim of discussing the effect of different physical and operating parameters on the column breakthrough time. The model used assumes LDF mass transfer kinetics; furthermore, a linear adsorption isotherm (i.e., $\lambda = 1$) will be considered. Nevertheless, all the conclusions that will be drawn would remain qualitatively unchanged if other mass transfer or isotherm models were used.

The solution of the system of partial differential equations obtained can be expressed in an analytical form [28]; nevertheless, such expression is quite complex and contains integrals that must be evaluated numerically. Therefore, the use of any other numerical method may be as demanding as using the "analytical" solution. Regardless of the method used to obtain the solution, here the attention will be focused on the effect of the model parameters on the shape of the breakthrough curve, in order to understand the column behavior.

By looking at the equations presented in Sect. 4.5.1.3, it appears that the solution depends on three parameters, namely μ, Pe, and St. A closer look reveals that this number can actually be reduced to two. Indeed, for any adsorption process worth being considered $\mu \gg 1$, therefore Eqs. 4.30 and 4.31 may be approximated as

$$\frac{\partial \tilde{q}}{\partial \tilde{t}} = -\frac{\partial \tilde{C}}{\partial \tilde{z}} + \frac{1}{Pe} \frac{\partial^2 \tilde{C}}{\partial \tilde{z}^2} \tag{4.35}$$

$$\frac{\partial \tilde{q}}{\partial \tilde{t}} = St \, \mu (\tilde{C} - \tilde{q}) \tag{4.36}$$

Equations 4.35 and 4.36 show that the solution of the model is significantly affected only by the value of the product $(St \, \mu)$ and Pe. By considering the definition of the Stanton number given in Sect. 4.5.1.3 and Eq. 4.21, the product $(St \, \mu)$ appears as a measure of the ratio between the rates of adsorption and column saturation.

Figure 4.5a shows the a dimensional breakthrough curves obtained by solving the above-presented model with different values of $(St \, \mu)$, while keeping $Pe = 100$. The order of magnitude of Pe was chosen by considering that, as noted earlier, for liquid mobile phases Pe_p is of the order of magnitude of 1 and

$$Pe = Pe_p \frac{d_c}{2R_p} \frac{L}{d_c}$$

where d_c is the column internal diameter. In order to avoid important wall channeling, the adsorbent particle diameter is generally chosen as to have $2R_p < 0.1 d_c$; furthermore, adsorption columns have often a high aspect ratio, so that it is not uncommon that L/d_c is of the order of magnitude of 10.

As the order of magnitude of the product $(St \, \mu)$ changes, different operating regimes of the adsorption column, corresponding to different features of the breakthrough curves, can be observed:

- Instantaneous adsorption $(St \, \mu \gg 1)$: Solute transfer to the solid phase is much faster than fluid convection rate, so that the local equilibrium condition holds. Similarly to an ideal column, the solute is completely removed from the mobile phase until the ideal column breakthrough time is approached. However, in this case, the breakthrough curve is smoothed by solute dispersion. The effect of dispersion on

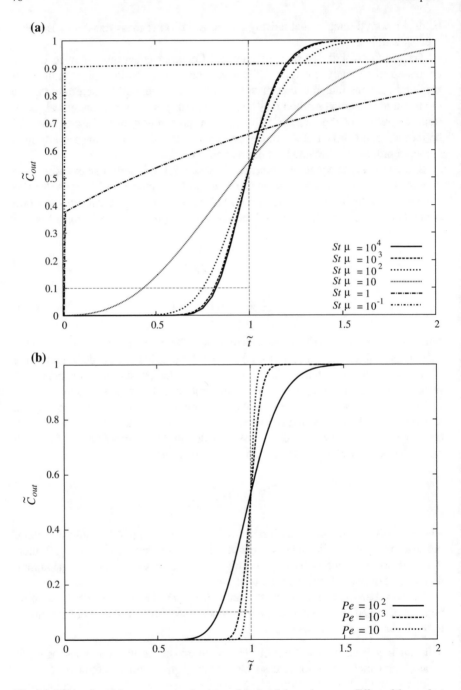

Fig. 4.5 Effect of model parameters on the shape of the breakthrough curve. **a** Effect of the product $St\,\mu$ $(Pe = 100)$. **b** Effect of Pe $(St\,\mu = 10^4)$

the breakthrough curve is shown in Fig. 4.5b, where it can be seen that as dispersion effects become less important (i.e., as Pe increases), the breakthrough curve approaches a step function, as expected for an ideal column.

- Fast adsorption rate ($St\,\mu \sim 10 - 10^2$): As the solute transfer rate decreases, the effects of mass transfer limitations become visible. The effect of slower mass transfer kinetics on the breakthrough curve is similar to an enhanced dispersion: The mass transfer zone widens, and the breakthrough curve becomes smoother. However, the column is still able to clear the solute from the fluid in the early operating time.

- Slow adsorption rate ($St\,\mu \sim 1$): If the mass transfer rate is too slow compared to the convection rate, the column can only lower the solute concentration without completely clearing it from the column effluent. Therefore, even in the early operating time, the outlet concentration is not zero.

- Negligible adsorption rate ($St\,\mu \ll 1$): If the solute transfer rate is negligible compared to fluid convection rate, no appreciable adsorption is observed and the solute concentration is unaffected by the passage through the column.

Clearly, well-designed fixed-bed adsorption columns should operate in the fast or instantaneous adsorption rate regimes.

An arbitrary threshold concentration of $0.1 C_{in}$ is marked with a dashed line in Fig. 4.5 to show the effect of Pe and St on the column breakthrough time. It is readily

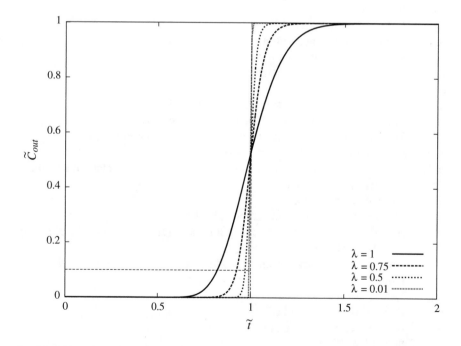

Fig. 4.6 Effect of the shape (linearity) of the adsorption isotherm on the breakthrough curve

seen that a decrease in one or both of these parameters can reduce significantly the column breakthrough time compared to what expected for an ideal column; therefore, mass transfer limitations and axial dispersion phenomena should be accounted for when designing a purification operation, especially if stringent constraints on the maximum admissible outlet concentration are imposed in the process.

In conclusion, in order to analyze the effect of the adsorption isotherm shape on the column breakthrough curve, the model presented in Sect. 4.5.1.3 has been solved with different values of λ while keeping $(St\,\mu) = 10^4$ (negligible mass transfer resistances) and $Pe = 10^2$. The results are shown in Fig. 4.6, where it can be seen that, all other conditions being equal, as the adsorption isotherm approaches the rectangular shape, the column breakthrough time increases.

Acronyms

LDF   Linear driving force
MTF   Mass transfer zone

Symbols

a_p	Specific interfacial area of solid particles
Bi_p	Particle Biot number
C	Concentration in the bulk of the liquid phase
C_s	Concentration in the liquid phase at the interface with the solid particles
d_0	Pore size
d_c	Column internal diameter
K_{lang}	Constant in the Langmuir adsorption isotherm (Eq. 4.3)
k_c	Mass transfer coefficient
k_{LDF}	LDF mass transfer coefficient
L	Column bed length
m	Partition coefficient for the linear adsorption isotherm (Eq. 4.5)
N_{fs}	Solute flux transferred from the fluid to the solid phase
$\langle N_z \rangle$	Surface-averaged axial solute flux on the column cross section
n_c	Number of components in the mixture
n	Specific adsorbed amount (amount of adsorbate per unit mass of adsorbent)
Pe	Modified Peclet number
Pe_p	Particle Peclet number
Q	Volumetric flow rate of the mobile phase
q	Adsorbate concentration on the solid phase (amount of adsorbate per unit volume of adsorbent)
q^*	Adsorbate concentration on the solid phase in equilibrium with the liquid phase
R_p	Radius of solid phase particles
Re_p	Particle Reynolds number
S	Column cross-sectional area
Sc	Schmidt number
St	Modified Stanton number

u	Superficial velocity of the liquid phase in the column
V_p	Volume of a solid particle
v	Interstitial velocity of the liquid phase in the column
v_f	Velocity of the adsorption front in the ideal column
ε	Void fraction of column bed
μ	Capacity ratio
μ_f	Viscosity of the fluid phase
λ	Adsorption isotherm linearity parameter
ρ_f	Density of the fluid phase
ρ_s	Density of the solid adsorbent
τ_{BT}	Breakthrough time
τ_{BT}^{*}	Breakthrough time of an ideal column
ζ	Axial position of the adsorption front in the ideal column
$\mathscr{D}_{dz}$	Axial dispersion coefficient
$\mathscr{D}_{mf}$	Molecular diffusion coefficient in the liquid phase
$\mathscr{D}_p$	Adsorbate effective diffusivity in solid particles

Subscripts

i	Related to component i
in	Column inlet
max	Maximum value
out	Column outlet
ref	Reference value

Overscripts

| $\sim$ | Dimensionless |
| $-$ | Volume average |

Chapter 5
Bioreactors

5.1 Introduction

A bioreactor is a system where chemical reactions involving biologically active substances, microorganisms, cells, or tissues are carried out. Bioreactors are widely used in chemical and biochemical industry processes such as fermentation, wastewater treatments, food processing, and production of recombinant proteins and pharmaceuticals (e.g., antibiotics and vaccines). In the field of biomedical applications, the main interest is devoted to bioreactors containing cells and tissues which are used to carry out biological functions lacking to natural organs to be supported or substituted. In particular, bioreactors are assuming a paramount role in tissue engineering, in which cells differentiated from stem cells are grown to obtain biological tissues.

Design and simulation of bioreactors for both for industrial and biomedical applications are based on common principles (mass and energy balances, biochemical and physical kinetic laws, etc.); however, biomedical reactors require stronger constraints in terms of operating conditions (temperature, pH etc.) and, furthermore, must ensure the maximum safety and be biocompatible. Designing a bioreactor is a very complex engineering task involving biological, biochemical and physical kinetics, fluid mixing, biomaterials, and control aspects. Therefore, modeling a bioreactor requires at least mass and energy balances, kinetics, and mass and heat transfer equations. The general model must be adapted to the specific type of bioreactor which has to be designed. A quite general form of mass balance in an infinitesimal volume element of incompressible fluid is the following one (see Chap. 1):

$$\frac{\partial C_i}{\partial t} = -\mathbf{u} \cdot \nabla C_i + \mathscr{D}_i \nabla^2 C_i + r_i \qquad (5.1)$$

where C_i is the concentration of component i, $\mathbf{u}$ the velocity of the fluid crossing the volume element, $\mathscr{D}_i$ the diffusivity of the component i, and r_i the specific reaction rate, i.e., the rate at which the component i is "generated" by chemical reaction per unit volume of fluid.

© Springer-Verlag London 2017
M.C. Annesini et al., *Artificial Organ Engineering*,
DOI 10.1007/978-1-4471-6443-2_5

Equation 5.1 shows that the variation rate of concentration C_i inside the control volume is given by the sum of three contributions: bulk flow of component i into the system, diffusion of i, and generation rate. When the balance is expressed in one-dimensional form, Eq. 5.1 becomes:

$$\frac{\partial C_i}{\partial t} = -u_x \frac{\partial C_i}{\partial x} + \mathscr{D}_i \frac{\partial^2 C_i}{\partial x^2} + r_i \tag{5.2}$$

A mass balance, as in Eq. 5.1 or 5.2, can be written for each component in the system. In many cases, Eqs. 5.1 and 5.2 simplify considerably. For example, if the system is working at steady state, the first term does not appear. Furthermore, if the composition is uniform inside the bioreactor, local balances are not required and algebraic equations can be used to express the balances of the whole reactor.

Likewise, the energy balance, coming from the first principle of thermodynamics, leads to the following expression:

$$\frac{\partial T}{\partial t} = -\mathbf{u} \cdot \nabla T + \frac{k_T}{\rho c_p} \nabla^2 T + \frac{r_i}{\rho c_p}(-\Delta H) \tag{5.3}$$

where T is the temperature, k_T the thermal conductivity, ρ the density, c_p the heat capacity, and ΔH the enthalpy of reaction. The terms in Eq. 5.3 are analogous to those appearing in Eq. 5.1. The first one represents the variation rate of temperature in unsteady state, the second and third terms are linked to flow of thermal energy by convective and diffusive mechanism, respectively, and the last term is the contribution due to the enthalpy of reaction. The heat generation rate is linked to the reaction rate r_i. In the one-dimensional form, Eq. 5.3 becomes:

$$\frac{\partial T}{\partial t} = -u_x \frac{\partial T}{\partial x} + \frac{k_T}{\rho c_p} \frac{\partial^2 T}{\partial x^2} + \frac{r_i}{\rho c_p}(-\Delta H) \tag{5.4}$$

Not all the terms appearing in the above equation are equally important; in some cases, the generation term is negligible compared with the other ones. Furthermore, in perfectly mixed reactors, differential balances can be conveniently substituted by a macroscopic balance.

The operation of bioreactors requires both heating and cooling systems devoted to supply heat at start-up, when working temperature has to be reached, and to remove the heat generated by the reactions. Because of the sensitivity of biological material to temperature, accurate control systems must be provided.

Both in mass and energy balances, the reaction rate plays a fundamental role, especially in biological reactions catalyzed by enzymes. The reaction rate r_i depends on the composition of the reacting system and on the environmental conditions, as temperature and pH. In the following section, some commonly used kinetic expressions are presented and shortly discussed.

5.2 Kinetics of Enzymatic Reactions

Most of biochemical reactions are catalyzed by enzymes that are macromolecules, usually proteins, able to transform reactant molecules (named substrates) at a very high reaction rate, often up to 10^{17} times faster than in absence of such biological catalysts. Furthermore, each enzyme is extremely specific so that it can quicken one and only one reaction without affecting the rate of many other reactions. These properties allow essential biological reactions to take place at rates required to sustain life. In vegetal and animal cells, thousands of reactions take place, most of which are catalyzed by enzymes; therefore, the cell itself can be considered as a very complex microbioreactor.

A biochemical process wherein enzymes catalyze the conversion of organic reagents to a desired product is called *fermentation*. Enzymes can be used free in solutions, or more frequently, they are entrapped or confined in solid matrices. Biomolecules as enzymes are very sensitive to temperature and pH so that an accurate control of these parameters is required during the fermentation process. Many enzymes (and cells) sharply lose their activity if the temperature changes more than 1–2 °C from the optimum value. In many cases, enzymes are not extracted from microorganisms containing them, but the whole cells are used as catalysts.

The most widely used kinetic expression of enzyme-catalyzed reactions is the Michaelis–Menten equation:

$$(-rs) = r_{\max} \frac{C_S}{k_M + C_S} \tag{5.5}$$

where $r_{\max}$ is the maximum reaction rate, k_M is the Michaelis constant, and C_S is the concentration of the reactant, called substrate.

Equation 5.5 shows a saturation trend which is typical of many biological processes. As a matter of fact, the reaction rate varies quite linearly with C_S at low concentrations ($C_S \ll k_M$), therefore the reaction can be considered of first order, whereas it tends to the constant value $r_{\max}$ at high values of C_S ($C_S \gg k_M$) and the order of reaction becomes zero.

The two parameters $r_{\max}$ and k_M are usually evaluated empirically by fitting experimental data. The maximum reaction rate $r_{\max}$ depends linearly on the concentration of the enzyme in solution:

$$r_{\max} = k(T) C_{E_A} \tag{5.6}$$

where C_{E_A} is the concentration of the active form of enzyme in solution. The kinetic constant $k(T)$ increases with temperature according to the Arrhenius law:

$$k(T) = k^0 \exp\left(-\frac{E_a}{RT}\right) \tag{5.7}$$

where the activation energy E_a have a magnitude in the range of 10–15 kcal/mole. However, temperature beyond the stability limit of the enzyme cause the denaturation of the protein with a drastic reduction of r_{max}. A simple model accounting for the dependence of r_{max} on temperature can be derived assuming a reversible reaction between the active (E_A) and the denatured (E_D) form of the enzyme:

$$E_A \leftrightarrows E_D \tag{5.8}$$

The equilibrium constant of reaction (5.8) can be expressed through the well-known equation:

$$\ln K_d = -\frac{\Delta G^0}{RT} = \frac{\Delta S^0}{R} - \frac{\Delta H^0}{RT} \tag{5.9}$$

where $\Delta G°$, $\Delta H°$, and $\Delta S°$ are the standard change of Gibbs free energy, enthalpy, and entropy of the deactivation reaction, respectively. The equilibrium constant depends on temperature according to the van't Hoff equation

$$\frac{d \ln K_d}{dT} = \frac{\Delta H^0}{RT} \tag{5.10}$$

The deactivation reaction reported in Eq. 5.8 is an endothermic process with $\Delta H° > 0$ (50–150 kcal/mole); therefore, the equilibrium constant K_d and, thus, the fraction of denatured enzyme increases with temperature. Through the definition of K_d in terms of concentrations of active and denatured forms, we get the following fraction f_A of active enzyme:

$$f_A = \frac{C_{E_A}}{C_{E,tot}} = \frac{1}{1 + K_d} = \frac{1}{1 + \exp\left(\dfrac{\Delta S^0}{R} - \dfrac{\Delta H^0}{RT}\right)} \tag{5.11}$$

where $C_{E,tot}$ is the total (i.e., active and denatured) enzyme concentration. As already pointed out, the maximum reaction rate is proportional to the active enzyme concentration; therefore, substitution of Eqs. 5.7 and 5.11 into Eq. 5.6 leads to the following form of the Michaelis–Menten equation:

$$(-rs) = k^0 \exp\left(-\frac{E_a}{RT}\right) \frac{1}{1 + \exp\left(\dfrac{\Delta S^0}{R} - \dfrac{\Delta H^0}{RT}\right)} C_{E,tot} \frac{C_S}{k_M + C_S} \tag{5.12}$$

Equation 5.12 shows that there is an optimal temperature where r_{max} (and therefore (-rs)) has a maximum value.

The active fraction of an enzyme in solution depends also on pH. Proteins are large molecules and may have many ionization states so that the charge on a protein in aqueous solution can be neutral, positive, or negative depending on the pH value of pH. The ionization state of an enzyme can affect the distribution of electric charge on its active site, possibly causing deactivation. A simple model to account for these phenomena can be based on the assumption that only two ionization states are possible (as for many amino acids). By considering the ionization equilibria involving the neutral (zwitterion) and charged forms of the enzyme

$$E^+ \overset{K_{C1}}{\leftrightarrows} E^{\pm} + H^+ \tag{5.13}$$

$$E^{\pm} \overset{K_{C2}}{\leftrightarrows} E^- + H^+ \tag{5.14}$$

it is possible to express the fraction f of each species as function of H^+ concentration:

$$f^{\pm} = \cfrac{1}{1 + \cfrac{[H^+]}{K_{C1}} + \cfrac{K_{C2}}{[H^+]}} \tag{5.15}$$

$$f^+ = \frac{[H^+]}{K_{C1}} f^{\pm} \tag{5.16}$$

$$f^- = \frac{K_{C2}}{[H^+]} f^{\pm} \tag{5.17}$$

where $[H^+]$ is the concentration of H^+ ions and K_{C1} and K_{C2} are the equilibrium constants of the reactions reported in Eqs. 5.13 and 5.14, respectively, that is,

$$K_{C1} = \frac{[H^+][E^{\pm}]}{[E^+]}$$

$$K_{C2} = \frac{[H^+][E^-]}{[E^{\pm}]}$$

In the case that only one of these forms is active, only its concentration should be accounted for in Eq. 5.6 and reaction rate equation becomes:

$$(-rs) = k f_i C_{E,tot} \frac{C_S}{k_M + C_S}.$$

5.3 Deviations from Michaelis–Menten Kinetics

The Michaelis–Menten equation, even though very widely used to represent the kinetics of enzymatic reactions, is not valid in many cases. A factor greatly affecting the reaction rate of enzymatic reactions is the presence of inhibitors, i.e., chemical species that interact with the enzyme and reduce its catalytic activity. Numerous

Table 5.1 Kinetic equations in the presence of inhibitors (adapted from [29])

Inhibition type	Reaction rate expression
Competitive	$(-rs) = r_{\max} \dfrac{C_S}{k_M \left(1 + \dfrac{C_I}{K_I}\right) + C_S}$
Uncompetitive	$(-rs) = r_{\max} \dfrac{C_S}{k_M + C_S \left(1 + \dfrac{C_I}{K_I}\right)}$
Non-competitive	$(-rs) = r_{\max} \dfrac{C_S}{(k_M + C_S) \left(1 + \dfrac{C_I}{K_I}\right)}$

examples of enzymatic process inhibition can be found in the biological and bio-medical fields. Sometimes, the effect of inhibitors can be positively exploited, as in the case of aspirin that inhibits the synthesis of prostaglandin involved in the pain-producing processes. In other cases, inhibitors can cause extremely detrimental effects, as in the case of cyanide, which inhibits the activity of cytochrome oxidase, thus stopping aerobic oxidation and causing death in a few minutes.

Sometimes, the same substrate or the product can inhibit the enzymatic reaction at high enough concentrations. Kinetic equations accounting for the presence of inhibitors have been proposed on the basis of mechanisms of reaction describing the interactions of the substrate and of the inhibitor with the enzyme molecule. Usually, three main types of mechanism are considered: competitive, uncompetitive, and non-competitive inhibition, even though several other mechanisms and kinetic expressions have been proposed. In competitive inhibition, substrate and inhibitor are very similar molecules that compete for the same active site on the enzyme molecule. Uncompetitive inhibition occurs when the inhibitor deactivates the complex formed by the bond between substrate and enzyme molecules. Finally, in the non-competitive mechanism, the inhibitor molecule attaches to an active site of the enzyme different from that one used by the substrate, but its presence deactivates the enzyme.

For each of the above-described mechanisms, a different expression of the reaction rate was derived. Such expressions are summarized in Table 5.1.

5.4   Mass Balance in a Well-Mixed Batch Enzymatic Reactor

A well-mixed reactor is a tank where composition and temperature are uniform in the whole volume of the reacting mixture. A batch reactor is a closed system where no mass flow is entering or exiting. Therefore, the first and second terms in the right-hand side of Eq. 5.2 are equal to zero and, in the case of enzymatic reactions, the following mass balance equation is obtained:

$$-\frac{dC_S}{dt} = r_{max}\frac{C_S}{k_M + C_S} \tag{5.18}$$

Equation 5.18 can be written in the dimensionless form:

$$\frac{d\tilde{C}_S}{d\tilde{t}} = -\frac{\tilde{C}_S}{\tilde{k}_M + \tilde{C}_s} \tag{5.19}$$

through the introduction of the following dimensionless variables and groups

$$\tilde{C}_S = \frac{C_S}{C_S^0} \quad \tilde{k}_M = \frac{k_M}{C_S^0} \quad \tilde{t} = \frac{t}{t^*} \quad t^* = \frac{C_S^0}{r_{max}} \tag{5.20}$$

Two limiting cases of zero-order and first-order approximations can be considered. When $\tilde{C}_S \gg \tilde{k}_M$ (zero-order approximation), Eq. 5.19 becomes

$$\frac{d\tilde{C}_S}{d\tilde{t}} = -1 \tag{5.21}$$

that results in a substrate concentration decreasing with a characteristic time t^*. When $\tilde{C}_S \ll \tilde{k}_M$ (first-order approximation), Eq. 5.19 becomes

$$\frac{d\tilde{C}_S}{d\tilde{t}} = -\frac{\tilde{C}_S}{\tilde{k}_M} \tag{5.22}$$

that results in a substrate concentration decreasing with a characteristic time

$$t^* = \frac{k_M}{r_{max}}. \tag{5.23}$$

5.5 Enzymatic Reactions with Entrapped Enzymes

In many industrial and biomedical applications, enzymes are not used as free molecules in solution but are attached to or confined to an inert and insoluble material. Immobilized enzymes can be easily removed from the reaction mixture and recycled as biocatalysts; furthermore, immobilization gives a greater stability to biomolecules even though their catalytic activity is often reduced by the presence of the solid support. Finally, in biomedical applications, the entrapment of enzymes within microcapsules, fibers, or gels does not cause an adverse response of the immunological system.

There are three main techniques for enzyme immobilization:

1. Adsorption on a solid support (such as glass or alginate beads),
2. Entrapment in porous beads or microspheres of calcium alginate, polyacrylamide etc.,
3. Cross-linkage, by which the enzyme is linked to the matrix by covalent bonds.

Each of these techniques presents some disadvantages. In adsorption, sometimes the active site of the enzyme is blocked by the adsorbent matrix with a dramatic reduction of the catalytic activity. The entrapment into porous particles requires the diffusion of substrate (and the counterdiffusion of products) across the pores. The consequent concentration gradient, sustained by the reaction, causes a reduction of the reaction rate in the inner zones of the particle. In cross-linkage technique, even though the binding site is different from the active site, the stiffness of the structure and the closeness of enzyme molecules sometimes preclude to substrate molecules to reach the active site of the enzyme. Use of spacer molecules (such as polyethylene glycol) placed between the enzyme molecules and solid support is a solution largely adopted to reduce the steric hindrance. Briefly, the following main criteria have to be taken into account in choosing the immobilization way:

• A high percentage of the enzyme must be retained in the matrix,
• The catalytic activity must be preserved, and
• The release of enzyme molecules must be prevented.

5.5.1 Kinetics of Enzymatic Reactions with Immobilized Enzymes

The presence of solid supports on which enzymes are fixed or entrapped brings about some important features which are not present when enzymes are free in a liquid solution. Firstly, the interaction with the solid surface may induce changes in the conformation of the immobilized enzymes or impair the accessibility of active sites; in both cases, the result can be a reduction of enzymatic activity. Furthermore, significant differences between the composition in the bulk of the liquid phase and surroundings of immobilized enzymes (where the reaction takes place) may be present. Such differences are mainly due to two physical phenomena: partition effects and mass transfer limitations.

Partition effects are caused by the microenvironment induced by the solid support, which is in general physicochemically different from the bulk of the liquid phase; thus, also at thermodynamic equilibrium, differences between the bulk and microenvironment composition may be present. Furthermore, due to diffusion limitations, substrate transport from the bulk solution to the reaction site is associated with a concentration gradient. More specifically, three different resistances can affect substrate transport:

- Mass transport resistance from the bulk solution to the solid support surface (external diffusion),
- Mass transport resistance across the membrane in the case of enzymes entrapped in microcapsules, hollow fibers etc., and
- Mass transport resistance inside the solid matrix entrapping enzyme molecules (internal diffusion).

The following part of this section shows how partition effects and diffusional limitations can be accounted for.

5.5.2 Partition Effects

Partition effects can be described through a *partition coefficient*, P, defined as follows:

$$P_i = \frac{C_i}{C_i^0} \tag{5.24}$$

where C_i is the concentration of the component i in the microenvironment and C_i^0 is the concentration of the same component in the bulk solution. The so-called *intrinsic kinetics* is determined by the local composition, but from an application point of view, it is more useful to express the reaction rate in terms of the bulk composition.

If Michaelis–Menten equation is used to express intrinsic kinetics and the partition coefficient of substrate can be assumed independent of composition, the following equation is obtained:

$$(-r_S) = \frac{r_{max} C_S}{k_M + C_s} = \frac{r_{max} P_s C_S^0}{k_M + P_s C_s^0} = \frac{r_{max} C_s^0}{\dfrac{k_M}{P_s} + C_s^0} \tag{5.25}$$

This equation describes an apparent kinetics where the Michaelis–Menten constant k_M is changed into k_M/P_s. The reaction rate increases or decreases in comparison with its intrinsic value depending on the value of P_S.

5.5.3 Effect of the External Diffusion

The mass flux of substrate due to external diffusion is (see Fig. 5.1)

$$J_s = k_{cs} \left(C_s^0 - C_{S,s} \right) \tag{5.26}$$

where C_S^0 and $C_{S,s}$ are the concentrations of substrate in the bulk and on the external solid surface, respectively, and k_{cs} is the transport coefficient of S. This coefficient

Fig. 5.1 External diffusion

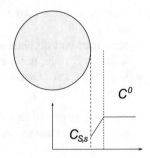

depends on the physical properties of the system and on the flow conditions around the particle. Its value can be evaluated through semiempirical relationships (see Chap. 2).

If enzyme molecules are bound on the external surface, the concentration at the reaction site is $C_{S,s}$. At steady state, the whole flow rate of substrate reaching the external surface is consumed by the reaction; therefore:

$$k_{sc}\left(C_S^0 - C_{S,s}\right) = \left(-r_S^s\right) = \frac{r_{max}^s C_{S,s}}{k_M + C_{S,s}} \tag{5.27}$$

In Eq. 5.27, $\left(-r_S^s\right)$ and r_{max}^s denote the reaction rate and its maximum value, respectively, per unit area of the external surface.

From Eq. 5.27, the unknown $C_{S,s}$ can be deduced and used to get the reaction rate in terms of operating variables. In terms of dimensionless variables, [30]

$$\tilde{C}_s = \frac{C_{S,s}}{C_S^0} \qquad \tilde{K}_M = \frac{K_M}{C_{S,s}^0} \qquad Da = \frac{r_{max}^s}{k_{Sc} C_S^0}$$

It is worth noting that the dimensionless group Da (Damköhler) is the ratio between the maximum reaction rate and the maximum mass transport rate.

Equation 5.27 becomes:

$$\frac{1 - \tilde{C}_S}{Da} = \frac{\tilde{C}_s}{\tilde{K}_M + \tilde{C}_s} \tag{5.28}$$

The presence of external diffusion makes the observed reaction rate different from the Michaelis–Menten values. In order to evaluate the effect of diffusion, an external effectiveness factor η_e can be defined as the ratio between the observed reaction rate and the reaction rate with no mass transport resistance, i.e., at a concentration C_S° on the external surface:

$$\eta_e = \frac{(-r_s)_{obs}}{\left[-r_s\left(C_S^0\right)\right]} \tag{5.29}$$

By expressing the reaction rate with the Michaelis–Menten expressions with $C_{S,s}$ as substrate concentration, Eq. 5.29 becomes:

$$\eta_e = \frac{\tilde{C}_s \left(\tilde{K}_M + 1 \right)}{\tilde{K}_M + \tilde{C}_s} \tag{5.30}$$

Since r_s is an increasing function of concentration and $C_{S,s} \leq C_{S,s}^0$, it can be stated that $(-r_s)_{obs} \leq -r_S \left(C_S^0 \right)$ and $\eta_s \leq 1$.

The weight of mass transport resistance on the observed reaction rate and on the effectiveness factor can be evaluated in terms of Damköhler number.

If $Da \rightarrow 0$ (i.e., if the maximum mass transport rate is much larger than the maximum reaction rate), $\tilde{C}_s \rightarrow 1$ (see Eq. 5.28) and the enzymatic reaction is the rate limiting step; in this case, the concentration of substrate on the solid surface is nearly equal to the concentration in the bulk so that the effectiveness factor becomes 1. On the contrary, if $Da \rightarrow \infty$ (i.e., if the maximum reaction rate is much larger than the maximum mass-transport rate), $\tilde{C}_s \rightarrow 0$ and diffusion in the external film is the rate-limiting step; the concentration on the solid surface becomes low enough to allow mass flux to balance the relatively high reaction rate. Accounting for Eq. 5.27, the observed reaction rate in this case is as follows:

$$(-r_s)_{obs} = k_{cs} C_S^0 \tag{5.31}$$

the effectiveness factor becomes 0 (see Eq. 5.30) (diffusion-limited regime). As shown in Eq. 5.30, η_e depends on $C_{S,s}$. However, in two limiting cases, a constant value of the effectiveness factor is obtained.

As $\tilde{C}_s \ll \tilde{K}_M$, Michaelis–Menten kinetics can be approximated by a first-order kinetics and Eq. 5.30 becomes:

$$\eta_e = \tilde{C}_s \tag{5.32}$$

With the assumption $\tilde{C}_s \ll \tilde{K}_M$, the term $\tilde{C}_s$ obtained from Eq. 5.28 is

$$\tilde{C}_s = \frac{\tilde{K}_M}{\tilde{K}_M + Da} \tag{5.33}$$

Substituting this result into Eq. 5.32 we get:

$$\eta_e = \frac{\tilde{K}_M}{\tilde{K}_M + Da} \tag{5.34}$$

Likewise, when $\tilde{C}_s \gg \tilde{K}_M$, Michaelis–Menten kinetics becomes zero order and both numerator and denominator in Eq. 5.29 are r_{max}^s so that we get $\eta_e = 1$. In this case, diffusion resistance, which reduces the concentration on the solid surface, has no effect since the reaction rate does not depend on the concentration.

5.5.4   Enzymes Entrapped in Microcapsules or Hollow Fibers

Enzymes can be dissolved in a liquid solution, which is separated from the substrate containing medium by means of a membrane. In this case, the solution with the enzyme molecules is separated from the bulk solution by a membrane permeable to the substrate molecules only.

Therefore, three resistances in series are present: the external resistance, the resistance of the membrane, and the resistance in the film adherent to the inner side of the membrane. In Fig. 5.2, partition effects between fluid and membrane material are shown as well. The enzymatic reaction occurs at $C_{S,i}$ substrate concentration.

If, for simplicity, the internal film resistance is assumed to be negligible, the following equations can be written at steady-state condition:

$$k_{cs}\left(C_s^0 - C_{S,s}\right) = P_m\left(C_{S,s} - C_{S,in}\right) = r_{max}^s \frac{C_{S,in}}{k_M + C_{S,in}} \qquad (5.35)$$

where the permeance P_m of the membrane depends on its thickness, on the effective substrate diffusivity in the membrane and on the substrate partition coefficient between the fluid phase and the membrane. Really, an additional term should be present in the flux through the membrane. It is associated with the possibility of the substrate to be dragged by the volume flux, and it is commonly expressed in terms of the so-called Stavermann coefficient, but in the present treatment, we assume negligible this term (see Chap. 3).

Fig. 5.2   Film and membrane resistances in a hollow fiber

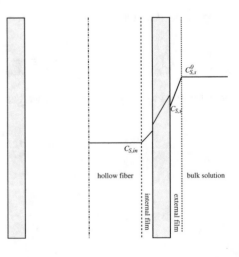

Equation 5.35 can be reformulated in terms of an overall resistance of transport phenomena in series:

$$\frac{C_s^0 - C_{S,i}}{R_{tot}} = r_{max}^s \frac{C_{S,i}}{k_M + C_{S,i}} \tag{5.36}$$

where

$$R_{tot} = R_{diff} + R_{membr} = \frac{1}{k_{sc}} + \frac{1}{P_m} \tag{5.37}$$

is the overall resistance to the transport phenomena. In terms of dimensionless quantities, we get:

$$\frac{1 - \tilde{C}_{S,in}}{Da_m} = \frac{\tilde{C}_{S,in}}{\tilde{K}_M + \tilde{C}_{S,in}} \tag{5.38}$$

where $\tilde{C}_{S,in} = C_{S,in}/C_s^0$, $\tilde{K}_M = k_M/C_s^0$, and Da_m is a modified Damköhler number defined as follows:

$$Da_m = \frac{R_{tot} r_{max}^s}{C_s^0} \tag{5.39}$$

An overall effectiveness coefficient can be defined also in this case as:

$$\eta_T = \frac{(-r_S)_{obs}}{\left[-r_s\left(C_S^0\right)\right]} = \frac{\tilde{C}_{S,i}\left(\tilde{K}_M + 1\right)}{\tilde{K}_M + \tilde{C}_{S,i}} \tag{5.40}$$

Following the same procedure as in Sect. 5.5.3, we can distinguish the two limiting cases in which the reaction approaches first-order and zero-order kinetics.

When $\tilde{C}_{S,i} \ll \tilde{K}_M$, first-order kinetics are approached. The overall effectiveness factor and $\tilde{C}_{S,in}$ can be evaluated from Eqs. 5.40 and 5.38, respectively, and the following result is obtained:

$$\eta_T = \tilde{C}_{S,in} = \frac{\tilde{K}_M}{\tilde{K}_M + Da_m} \tag{5.41}$$

which is formally identical to Eq. 5.34. Also in this case, if $Da_m = 0$, the overall effectiveness factor is 1, since the overall resistance is negligible and $C_{S,in} = C_S^0$.

However, intermediate cases are possible. Only if the external diffusion resistance is negligible, Damköhler number becomes

$$Da_m = \frac{R_{membr} r_{max}^s}{C_S^0} = \frac{r_{max}^s}{P_m C_S^0} \tag{5.42}$$

and at low values of P_m, the process is controlled by diffusion across the membrane.

Likewise, if the membrane resistance is negligible in comparison with the external diffusion resistance, Da_m becomes equal to Da and $\eta_T = \eta_e$.

When $\tilde{C}_{S,in} \gg \tilde{K}_M$, zero-order kinetics are approached. As already shown in Sect. 5.5.3, in this case $\eta_T = 1$.

5.5.5 Internal Diffusion

As in non-enzymatic heterogeneous catalysis, a technique typically used in reactions catalyzed by enzymes is based on immobilizing them on the internal surface of porous supports permeable to substrate molecules. In this case, the diffusion of substrate through the matrix is coupled with the reaction so that the substrate concentration decreases from the external surface toward the inner zones of the porous support. As a consequence, the local reaction rate decreases as well. Since the design of a bioreactor is based on the observed reaction rate, i.e., on the rate by which the substrate is consumed in the whole volume of the solid support, some considerations are required in order to find the observed reaction rate from substrate concentration profile inside the support.

Assuming the porous support entrapping the enzyme as a pseudo-homogeneous structure with a spherical shape, the substrate concentration profile can be derived from the local mass balance equation, as reported in Chap. 1. Here, a differential shell defined by two concentric spheres of radius r and $r + dr$ (Fig. 5.3) is considered as control volume and the mass balance equation is written as:

$$\left(\mathscr{D}_{e,S} \frac{dC_S}{dr} 4\pi r^2 \right)\bigg|_{r+dr} - \left(\mathscr{D}_{e,S} \frac{dC_S}{dr} 4\pi r^2 \right)\bigg|_{r} = (-r_S)\, 4\pi r^2 dr \qquad (5.43)$$

Fig. 5.3 Spherical shell of thickness dr: control volume for the mass balance reported in Eq. 5.43

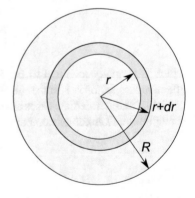

where $(-r_S)$ is the reaction rate per unit volume of particle and $\mathscr{D}_{e,S}$ is the effective diffusion coefficient of substrate that accounts for the porosity of the particle and the tortuosity of pores.

From the above mass balance, the following equation is obtained:

$$\frac{d^2 C_s}{dr^2} + \frac{2}{r}\frac{dC_s}{dr} = \frac{(-r_S)}{\mathscr{D}_{e,S}} \tag{5.44}$$

In terms of the following dimensionless quantities:

$$\tilde{C}_s = \frac{C_s}{C_S^0} \quad \tilde{r} = \frac{r}{R} \quad \beta = \frac{C_S^0}{k_M} \tag{5.45}$$

and assuming the Michaelis–Menten expression for the reaction rate, Eq. 5.44 becomes:

$$\frac{d^2 \tilde{C}_S}{d\tilde{r}^2} + \frac{2}{\tilde{r}}\frac{d\tilde{C}_S}{d\tilde{r}} = \Phi^2 \frac{\tilde{C}_S}{1 + \beta \tilde{C}_S} \tag{5.46}$$

The dimensionless number Φ (Thiele modulus) is defined as follows:

$$\Phi = R\sqrt{\frac{r_{max}/k_M}{\mathscr{D}_{e,S}}} \tag{5.47}$$

and its value gives information about the ratio between the diffusion and reaction characteristic times.

In defining the dimensionless quantities $\tilde{C}_S$ and β, the substrate concentration C_S^0 in the bulk has been used. Another possibility should be using the substrate concentration on the surface of the particle $C_{S,s}$. Obviously, these two quantities have the same value if the external transport resistance is negligible.

Equation 5.46 can be solved with the following boundary conditions holding at negligible external transport resistance:

$$\tilde{C}_S\Big|_{\tilde{r}=1} = 1 \qquad \frac{d\tilde{C}_S}{d\tilde{r}}\Big|_{\tilde{r}=0} = 0 \tag{5.48}$$

to get the substrate concentration profile and therefore an effectiveness factor, defined as the ratio between the *observed* reaction rate and the reaction rate calculated as if all the particle were at the substrate concentration on the external surface. Since, at steady-state condition, the observed substrate consumption rate is equal to its flow rate through the external surface of the particle, the effectiveness factor can be written in the following form:

$$\eta = \frac{4\pi R^2 \mathcal{D}_{e,S} \left.\dfrac{dC_S}{dr}\right|_R}{\dfrac{4}{3}\pi R^3 \dfrac{r_{max}C_S^0}{k_M + C_S^0}} = \frac{3}{R} \frac{\mathcal{D}_{e,S} \left.\dfrac{dC_S}{dr}\right|_R}{\dfrac{r_{max}C_S^0}{k_M + C_S^0}}$$

(5.49)

or in terms of dimensionless quantities:

$$\eta = 3 \frac{\left(\dfrac{d\tilde{C}_S}{d\tilde{r}}\right)_{\tilde{r}=1}}{\Phi^2 \left(\dfrac{1}{1+\beta}\right)}$$

(5.50)

Equation 5.46 is a second-order nonlinear differential equation and no analytical solution can be found. In this general condition, only a numerical solution can be found both for determining the concentration profile and for evaluating η. However, in two simpler approximations, analytical solutions can be obtained.

5.5.5.1 First-Order Reaction Approximation

If $C_S^0 \ll k_M$, or $\beta \ll 1$, Eq. 5.46 becomes linear and the following concentration profile is obtained:

$$\tilde{C}_S = \frac{1}{\tilde{r}} \frac{\sinh(\Phi_1 \tilde{r})}{\sinh \Phi_1}$$

(5.51)

where $\Phi_1 = \Phi$ is the Thiele modulus for a first-order reaction. By using this result, the effectiveness factor defined in Eq. 5.50 becomes [29]:

$$\eta = \frac{3}{\Phi_1^2} (\Phi_1 \coth \Phi_1 - 1)$$

(5.52)

Therefore, in the first-order approximation, the effectiveness factor depends only on the Thiele modulus. Figure 5.4 reports a plot of the effectiveness factor as a function of the Thiele modulus according to Eq. 5.52. At low values of Φ_1, the effectiveness factor is nearly 1 and the reaction is surface reaction limited; on the contrary, η becomes very small when the Thiele modulus is large and the reaction is diffusion-limited within the particle. For very large, Φ_1, Eq. 5.52 can be approximated by $\eta \approx 3/\Phi_1$.

Since values of η close to 1 make the observed reaction rate close to its maximum value, it is advisable to operate with a low value Φ_1, e.g., with low-diameter solid particles.

Fig. 5.4 Dependence of η on Φ for a first-order reaction in spherical particles

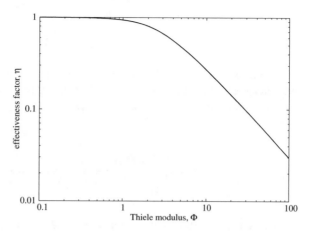

5.5.5.2 Zero-Order Reaction Approximation

In the case $C_S^0 \gg k_M$ and $(-r_S) = r_{\max}$, Eq. 5.44 becomes:

$$\frac{d^2 C_s}{dr^2} + \frac{2}{r} \frac{dC_s}{dr} = \frac{r_{\max}}{\mathcal{D}_{e,S}} \tag{5.53}$$

In this case, the substrate concentration profile is given by:

$$C_S = C_S^0 - \frac{r_{\max}}{6\mathcal{D}_{e,S}} \left(R^2 - r^2\right) \tag{5.54}$$

or in terms of dimensionless quantities:

$$\tilde{C}_S = 1 - \frac{\Phi_0^2}{6} \left(1 - \tilde{r}^2\right) \tag{5.55}$$

where Φ_0 is the Thiele modulus for a zero-order reaction:

$$\Phi_0 = R \sqrt{\frac{r_{\max}}{\mathcal{D}_{e,S} C_S^0}} \tag{5.56}$$

Unlike Φ_1, the Thiele modulus in this case depends on the substrate concentration, C_S^0, on the external surface of the particle.

Equations 5.54 and 5.55 hold as long as C_S or $\tilde{C}_S$ are nonnegative, i.e., up to the radial position r_C where the substrate is exhausted. The value of $\tilde{r}_C = r_C/R$ can be easily found by setting $\tilde{C}_S = 0$ in Eq. 5.55:

$$\tilde{r}_C = \sqrt{1 - \frac{6}{\Phi_0^2}} \tag{5.57}$$

If Eq. 5.57 admits a real root $0 \leq \tilde{r}_C < 1$, then the reaction takes place only in the outer shell of the particle $\tilde{r}_C < \tilde{r} \leq 1$, whereas in the core $\tilde{r} < \tilde{r}_C$, the substrate concentration and reaction rate are zero[1]; in this case, the effectiveness factor can be calculated as

$$\eta = \frac{\frac{4}{3}\pi \left(1 - \tilde{r}_C^3\right) r_{max}}{\frac{4}{3}\pi r_{max}} = 1 - \tilde{r}_c^3 = 1 - \left(1 - \frac{6}{\Phi_0^2}\right)^{3/2} \tag{5.58}$$

Differently, if Eq. 5.57 does not admit real solution, the reaction takes place throughout the particle and the effectiveness factor is 1.

Finally, if the resistance to mass transport in the external film is not negligible, substrate concentration on the external surface of the pellet can be found by equating the flux through the film to the diffusion flux entering into the solid.

5.6 Bioreactor for Cell Cultures

Culturing cells have many points in common with enzymatic reactions, but further, very important specific requirements have to be taken into account. For example, cells are very fragile and shear sensitive; furthermore, they frequently require an anchorage to a surface and are therefore grown on microcarriers or on scaffolds. Finally, many biological functions are not performed if cells are not in contact each other as in a tissue.

Bioreactors of major interest in biomedical field are those ones where animal cells are cultivated in an environment attempting to mimic physiologic conditions, as much as possible.

Depending on the type of the bioreactor used, on the supply of nutrients and on the stage of the cell culture, cell population can grow in the time or can remain in a steady state where it carries out its typical physiological functions. Supplying essential nutrients at the rate required from the cells is fundamental in design and operation of such bioreactors. A typical example of this difficult task is the supply of oxygen in aerobic processes because of the poor solubility of O_2 in water, which requires pressurization, and of the need of agitation for helping oxygen transfer with the consequent risk of damaging the cells. Besides limiting nutrients, in tissue engineering applications, a continuous stimulation of cells by adequate mechanical,

---

[1]It is noteworthy that the zero-order approximation holds only at $C_S \gg k_M$ so that, when the substrate concentration is low enough at the inner core, a transition occurs from zero-order to first-order reaction rate.

electrical, and chemical signals is required in order to keep cells viable, differentiated, and organized.

Although some cell cultures can be grown suspended in solution (e.g., hybridomas), most cells used for bio-artificial organs or in tissue engineering are anchorage-dependent cells, i.e., cells requiring a surface such as a plastic solid support (scaffold) or a microcarrier. Sometimes, cells are entrapped in gel beads likewise as for enzymes. The entrapment produces some problem in mass transport but protects the cells from mechanical stresses generated by the agitation. A further non-negligible problem in the operation of a bioreactor containing cells arises from the need of overall sterility of the equipment in order to avoid possible contamination of the culture.

Following the same approach used for the enzymatic reactions, the next sections provide an overview of the kinetics of cell growth and substrate utilization in suspended cultures; then, some remarks on the mass transport limitations in entrapped cells are presented, together with simple models useful in describing the behavior of supported or entrapped cells.

5.6.1 Kinetics of Cell Growth and of Product Formation

Several more or less sophisticated approaches are possible in describing cell populations. At the simplest level, it can be considered as a homogeneous population (biomass) of single undifferentiated units. Models based on this assumption are called *unsegregated and unstructured models*.

More sophisticated approaches take into account the cell population heterogeneity due to the distribution of one or more properties (type and age of cells, rate of synthesis of the desired products etc.). Models accounting for such differences are called *segregated models*. Furthermore, each cell can be described through its internal structure (multicompartment) in the attempt of accounting for metabolic pathways. The corresponding models are called *structured models*. Obviously, the more complex is the model, the more involved is its solution and application to the analysis of bioreactors. Furthermore, the possibility of constructing reliable complex models is closely linked to the level of biological knowledge. A thorough presentation of cell population models can be found in specialized texts [30]; here, only a simple approach is presented.

In order to analyze a biological process involving biomass, the starting point is the knowledge of the rates of nutrient consumption, product generation rate, and, especially, of the biomass growth. These different rates are related to each other through the stoichiometry of the reactions that take place inside the cells, and in a simple approach, they can be expressed by means of *yield factors*. Since the growth of biomass is related to cell multiplication, its rate (per unit volume, for instance) can be set proportional to the biomass concentration X:

$$r_X = \mu X \tag{5.59}$$

where μ indicates the specific growth rate. This quantity, which has the units of inverse time, depends on the nutrient concentration in the culture medium. A simple dependence of μ on the concentration of one essential nutrient (the concentration of all other components being constant) was proposed by Monod:

$$\mu = \mu_{\max}\frac{C_S}{k_S + C_S} \tag{5.60}$$

Equation 5.60 has the same form of the Michaelis–Menten equation, in accordance with the observation that cell growth is a consequence of the enzymatic reactions occurring in the cell. Even though the Monod model can be considered as an over-simplification, it is widely used because it has a reasonable consistency and requires only two empirical parameters: $\mu_{\max}$ and k_S. For most organic substrates, k_S values range from 0.1 to 20 mg/L and $\mu_{\max}$ takes values ranging from 0.3 to 1 h^{-1}. The specific growth rate is greatly affected by temperature and pH, since the growth of the cell population depends on the enzymatic mechanisms occurring inside the cells. Temperature has a positive effect on the growth rate up to a value beyond which the denaturation of many enzymatic proteins causes a rapid decline of the growth rate and cell death. It is worth to remark that, in many cases, the temperatures that maximize cell growth rate and yield of desired products do not coincide. Likewise, the pH of the medium is another important factor for the culture of cells both for the effect on the cell metabolism and for the state of dissociation of the molecules in the culture medium.

As in the Michaelis–Menten equation, also in Monod model, two approximations are possible for Eq. 5.60 when $C_S \ll k_S$ (first-order kinetics) or $C_S \gg k_S$ (zero-order kinetics).

Beside Monod equation, several other kinetic expressions were proposed for the dependence of μ on C_S; three of the most commonly used equations are reported in Table 5.2. The Tessier and Moser equations show a trend similar to Monod equation, but are formally more complex; furthermore, the Moser equation has three empirical parameters instead of two. The Contois equation is based on the "unbalanced growth" approach; i.e., μ depends on the biomass concentration so that the specific growth rate decreases as the cell population increases.

Furthermore, deviations from the Monod kinetic law are sometimes observed when inhibition effects by substrate or product are present. Two equations are often used for substrate and product inhibition, respectively:

Table 5.2 Commonly used expressions for cell growth rate

Author	Equation
Tessier	$\mu = \mu_{\max}\left[1 - \exp\left(-C_S/k_S\right)\right]$
Moser	$\mu = \mu_{\max}\left[1 + k_S C_S^{-\lambda}\right]^{-1}$
Contois	$\mu = \mu_{\max}C_S \left(B \cdot X + C_S\right)^{-1}$

$$\text{substrate inhibition} \quad \mu = \mu_{\max} = \frac{C_S}{k_S + C_S + \dfrac{C_S^2}{K_I}} \tag{5.61}$$

$$\text{product inhibition} \quad \mu = \mu_{\max} \frac{C_S}{(C_S + k_S)\left(1 + \dfrac{C_P}{K_P}\right)} \tag{5.62}$$

where C_P is the product concentration and K_I and K_P are parameters related to the inhibition mechanism.

The above-presented equations provide the growth rate of biomass. However, the biomass can attain a steady state, in which population growth is not observed and the nutrient is used only for the maintenance of vital functions of the cells. This is the case of cells used in bio-artificial organs during the stage of medical treatment.

In designing a bioreactor, mass balance of products is usually required as well. The simplest way to give a kinetic expression for the product generation is to link this quantity to the rate of substrate consumption or to the behavior of the cell population. Two kinds of product generation can occur: one is associated with the growth of the cells and the other one is non-growth associated. In the first mechanism, the production rate is proportional to the biomass growth rate, whereas in the second one, it is proportional to the cell concentration. A model proposed by Leudeking and Piret [31] combines both the terms in a kinetic equation containing two empirical parameters (α and β):

$$r_P = \alpha \mu + \beta X \tag{5.63}$$

5.6.2 Kinetics of Cell Cultures with Immobilized Cells

A long-term preservation of metabolic functions of many cells requires their adhesion to suitable solid matrices and their culture at high cell density. These requirements make diffusion of nutrients and of cell products a critical point in culturing cells, especially in three-dimensional structures. In order to face this kind of problems, mathematical models are very useful. In the following two case studies are presented and discussed through the use of simple models.

5.6.2.1 Hepatocytes Entrapped in Alginate Gel Beads

In the lots of attempts of developing bioreactors for a bio-artificial liver, the entrapment of hepatocytes in calcium alginate gel beads has proved to be a promising immobilization technique. Biocompatibility, chemical stability, and permeability of the polymer support meet some of the requirements for the use in a bio-artificial organ. Furthermore, cells entrapped in alginate gel are able to organize themselves in a liver-like three-dimensional structure.

However, the critical point of this technique is the supply of nutrients and especially of oxygen to the cells located at the inner part of the beads. Low oxygen concentrations ($< 4 \times 10^{-2}$ mM) cause unacceptable changes in the metabolic functions of hepatocytes and can eventually result in death of cells. Oxygen molecules diffuse from the outer surface of the bead toward the center of the particle, while it is consumed by the cells met in the path. Therefore, a concentration profile is set inside the bead. The characteristics of this profile depend on the O_2 concentration on the external surface (and thus on the concentration on the bulk liquid), on the cell density, and on the size of the particle.

A reaction–diffusion model [32] is very useful to study the conditions that allow anoxic zones to be avoided anywhere in the particle. The model is based on the following assumptions:

- spherical particles of radius R;
- homogeneous distribution of the cells in the bead;
- uniform O_2 effective diffusivity within the particle;
- negligible external mass transfer resistance;
- pseudo-zero-order O_2 consumption rate (OCR) by the cells at O_2 concentration greater that a fixed threshold C_{O2}^c;
- OCR $= 0$ where O_2 concentration is below the threshold because of anoxia of hepatocytes.

The latter two assumptions can be represented by the following stepwise function:

$$\begin{cases} C_{O2} \leq C_{O2}^c & (-r_{O2}) = 0 \\ C_{O2} > C_{O2}^c & (-r_{O2}) = q \cdot n_V' \end{cases} \tag{5.64}$$

where q is the OCR per cell and $n_V' = n_V / V$ is the density of viable cells, i.e., the number n_V of viable cells entrapped in the volume V_B of the bead.

The steady-state mass balance of oxygen in a shell of thickness dr inside the bead is as follows (see Fig. 5.3):

$$\frac{d}{dr}\left(r^2 \frac{dC_{O2}}{dr}\right) = \frac{q \cdot n_V'}{\mathscr{D}_{e,O2}} \cdot r^2 \tag{5.65}$$

where $\mathscr{D}_{e,O2}$ is the effective O_2 diffusivity in the gel matrix. Equation 5.65 is analogous to equation derived for zero-order enzymatic reactions (see Eq. 5.53), but it holds only where $C_{O2} > C_{O2}^c$, i.e., for $r_c \leq r \leq R$. Obviously, the critical radius r_c can be also be equal to 0, if the whole particle is well oxygenated.

In terms of dimensionless variables $\tilde{C}_{O2} = (C_{O2}/C_{O2,s})$ and $\tilde{r} = (r/R)$, Eq. 5.65 can be cast in the following dimensionless form:

$$\frac{d}{dr}\left(\tilde{r}^2 \frac{d\tilde{C}_{O2}}{d\tilde{r}}\right) = \left(R^2 \frac{q \cdot n_V'}{\mathscr{D}_{e,O2}C_{O2,s}}\right)\tilde{r}^2 = \Phi^2 \tilde{r}^2 \tag{5.66}$$

where

$$\tilde{C}_{O2} = \frac{C_{O2}}{C_{O2,s}}; \quad \tilde{r} = \frac{r}{R} \tag{5.67}$$

and

$$\Phi = R\sqrt{\frac{q \cdot n'_v}{\mathscr{D}_{e,O2}C_{O2,s}}} \tag{5.68}$$

is the Thiele modulus for a zero-order reaction.

The boundary conditions for the integration of Eq. 5.66 are as follows:

$$\tilde{r} = 1 \quad \tilde{C}_{O2} = 1 \quad \text{and} \quad \tilde{r} = \tilde{r}_c \quad \frac{d\tilde{C}_{O2}}{d\tilde{r}} = 0 \tag{5.69}$$

where $\tilde{r}_c = (r/r_c)$. The solution of Eq. 5.66 gives the following O_2 concentration profile

$$\tilde{C}_{O2} = 1 - \frac{1}{6}\Phi^2\left(1 - \tilde{r}^2\right) - \frac{1}{3}\Phi^2\tilde{r}_c^3\left(1 - \frac{1}{\tilde{r}}\right) \tag{5.70}$$

The term $\tilde{r}_c^3$ locates the anoxic volume fraction. Since $\tilde{C}_{O2} = \tilde{C}_{O2}^c$ at $\tilde{r} = \tilde{r}_c$, the following equation can be deduced

$$\tilde{r}_c^3 - \frac{3}{2}\tilde{r}_c^2 + 3\frac{\tilde{C}_c - 1}{\Phi^2} - \frac{1}{2} = 0 \tag{5.71}$$

The solution of this equation allows $\tilde{r}_c$ to be calculated in terms of the Thiele modulus and the critical O_2 level. Figure 5.5 shows the dependence of the anoxic dimensionless radius $\tilde{r}_c$ on the Thiele modulus when $\tilde{C}_c = 0.005$. For $\Phi < 2.44$, O_2 concentration is above the threshold value throughout the spherical particle, while for $\Phi > 2.44$, an anoxic core is present whose radius increases with Φ. Obviously, for a given value of Φ, the oxygen concentration threshold $\tilde{C}_c$ affects the volume of the anoxic zone: As

Fig. 5.5 Anoxic radius for $\tilde{C}_c = 0.005$

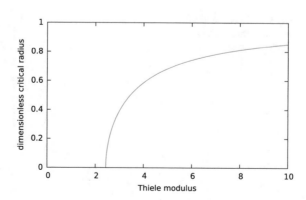

an example, if $\tilde{C}_c = 0.5$, the maximum allowable value of Φ to avoid the formation of an anoxic core is 1.73. In general, in order to ensure a complete oxygenation of the spherical particle, the following condition must be satisfied:

$$\Phi^2 = 6\left(1 - \tilde{C}_c\right) \tag{5.72}$$

It is worthwhile to observe that in the artificial liver, a high cell density n'_V is required. Therefore, a small enough Thiele modulus can be achieved using small particles (small R), even though this solution produces large pressure drops in packed bed bioreactors.

Figure 5.6 shows O_2 concentration profiles Eq. 5.70 at three values of Φ for $\tilde{C}_c = 0.005$.

5.6.3 Immobilization of Cells on Microporous Membranes

Another technique used in bio-artificial liver is based on the immobilization of the cells on the inner or outer surface of hollow fibers made of a porous material. This configuration allows oxygen and other nutrients to be supplied to cells by means of a liquid stream different from the blood or plasma stream. The nutrient stream flows in the membrane side where cells are fixed, whereas blood plasma flows in the other side. The flow scheme can be cocurrent or countercurrent, but usually the cocurrent scheme is adopted. The module is a bundle of hollow fibers contained in a cylindrical shell. A main problem is the level of O_2 concentration on the cells at the outlet of the feed stream; actually, this section is the most critical, due to oxygen consumption along the fiber from the inlet to the outlet section. To this purpose, a model is presented for calculating radial and axial distributions of O_2 concentrations in order to find possible critical zones for the cell viability. The configuration to be modeled is based on the following assumptions:

Fig. 5.6 O_2 concentration profiles at three values of Thiele modulus

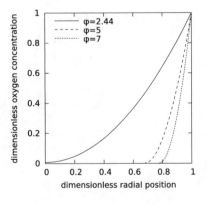

- Nutrients flow in the extrafiber space,
- Cells are immobilized on outer surface of the fibers,
- Plasma flows through the lumen of fibers.

Nutrient stream flows cocurrently with plasma stream in order to ensure a higher oxygen concentration where a higher metabolic functionality is required, that is, for the cells located at the inlet section where the concentration of plasma toxins is higher. At any cross section, oxygen transfer from nutrient stream toward cells occurs through the following steps:

- Mass transport in the external fluid film;
- Diffusion with reaction through layers of cell population covering the fiber;
- Mass transport through the membrane;
- Mass transport in the plasma fluid flowing inside the capillary fiber.

In modeling the device, membrane resistance is assumed to be negligible because of its low thickness and its high cutoff for oxygen molecules. The model is based on the following assumptions:

- Steady-state conditions;
- Constant volumetric flow rates along the axis of the device;
- Plug flow of nutrient stream;
- Homogeneity of cell layer stuck to fibers: Oxygen diffuses radially though the layers and is consumed because of cell respiration and other metabolic reactions;
- Laminar flow of plasma stream in the lumen: The concentration of solutes changes both with radial and axial position. Mass transport mechanism is diffusive along the radius and convective along the axis;
- Plasma is assumed to be a Newtonian fluid.

The mathematical model is based on local mass balances of oxygen in the various control volumes (Fig. 5.7), together with transport and reaction rate equations.

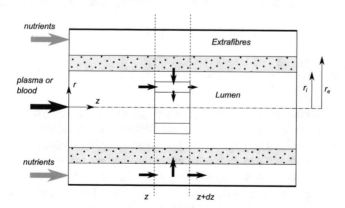

Fig. 5.7 Control volumes for mass balances in a single fiber

The mass balance in the extrafiber volume can be written as:

$$\frac{dC_L(z)}{dz} + \frac{2\pi r_e k_e n_f}{Q_L}[C_L(z) - C(r_e, z)] = 0 \tag{5.73}$$

where $C_L(z)$ and $C(r_e, z)$ are oxygen concentrations in the bulk fluid and on the external cell layer at axial position z, respectively; Q_L is the volumetric flow rate of the nutrient fluid; k_e is the oxygen transport coefficient; and n_f is the number of fibers. The boundary condition is $C_L(0) = C_L^0$ at $z = 0$, where C_L^0 is the oxygen concentration in the nutrient stream entering the module.

In the cell layer, oxygen transport is coupled with metabolic reaction of cells. Assuming a zero-order reaction, mass balance in an infinitesimal volume long dz and with thickness dr within this layer is as follows:

$$\mathscr{D}_{eff}\left(\frac{\partial C(r, z)}{\partial r} + r\frac{\partial^2 C(r, z)}{\partial r^2}\right) = q \cdot n_V' \cdot r \tag{5.74}$$

where $\mathscr{D}_{eff}$ is the oxygen effective diffusivity in the cell layer and q is the oxygen consumption rate per cell. The cell density n_V' can be evaluated from the total number of cells n_V in the bioreactor:

$$n_V' = \frac{n_V}{\pi\left(r_e^2 - r_i^2\right)L n_f} \tag{5.75}$$

where L indicates the fiber length. By substituting Eq. 5.75 in Eq. 5.74, we get:

$$\mathscr{D}_{eff}\left(\frac{\partial C(r, z)}{\partial r} + r\frac{\partial^2 C(r, z)}{\partial r^2}\right) = q\frac{N}{\pi\left(r_e^2 - r_i^2\right)L n_f}r \tag{5.76}$$

Equation 5.76 can be integrated with the following boundary conditions:

$$C(z) = C_i(z) \quad \text{at} \quad r = r_i \quad \forall z \tag{5.77}$$

$$\mathscr{D}_{eff}\left.\frac{\partial C(r, z)}{\partial r}\right|_{r=r_e} = k_e[C_L(z) - C(r_e, z)] \quad \text{at} \quad r = r_e \quad \forall z \tag{5.78}$$

In the intralumen volume, the microscopic mass balance can be derived by considering a hollow cylinder with length dz and thickness dr, located inside the internal fluid and coaxial with the fiber (Fig. 5.7). Oxygen enters this volume by convective transport with the liquid stream flowing along the fiber at z position and by diffusive transport through the cylinder side surface with radius $(r + dr)$. Then, oxygen exits from the volume by the same mechanisms, at $(z + dz)$ and through the inner cylinder of radius r. Since the liquid flow is laminar, the velocity pattern is parabolic in shape with a maximum velocity at the center equal to two times the average velocity:

$$u_z(r) = 2 \frac{Q_P}{\pi r_i^2 n_f} \left[1 - \left(\frac{r}{r_i} \right)^2 \right] \tag{5.79}$$

In this equation, Q_P indicates the overall plasma volumetric flow rate flowing through the whole bundle of fibers. The oxygen balance equation is as follows:

$$u_z(r) C_P(r,z) dS_b|_z + J \cdot dS_l|_{r+dr} = u_z(r) C_P(r,z) dS_b|_{z+dz} + J \cdot dS_l|_r \tag{5.80}$$

where C_P indicates the oxygen concentration in the plasma, J is the diffusive flux, and dS_b and dS_l are the areas of base and side surfaces, respectively, of the hollow cylinder.

If the diffusive flux is expressed by Fick equation, the mass balance becomes:

$$2 \frac{Q_P}{\pi r_i^2 n_f} \left[1 - \left(\frac{r}{r_i} \right)^2 \right] \cdot r \frac{\partial C_P(r,z)}{\partial z} = \mathscr{D}_P \left(\frac{\partial C_P(r,z)}{\partial r} + r \frac{\partial^2 C_P(r,z)}{\partial r^2} \right) \tag{5.81}$$

with the boundary conditions:

$$\mathscr{D}_P \frac{\partial C_P(r,z)}{\partial r} = 0 \quad \text{at} \quad r = 0 \quad \forall z \tag{5.82}$$

$$\mathscr{D}_P \frac{\partial C_P(r,z)}{\partial r} = \mathscr{D}_{eff} \frac{\partial C(r,z)}{\partial r} \quad \text{at} \quad r = r_i \quad \forall z \tag{5.83}$$

$$C_P(0,0) = C_P^0 \quad \text{at} \quad z = 0; \quad r = 0 \tag{5.84}$$

Last equation holds when plasma entering the bioreactor is oxygenated. If plasma is not oxygenated, the inlet O_2 concentration has to be set 0.

Solving the set of differential Eqs. 5.73, 5.76, and 5.81 with the corresponding boundary conditions allows to determine the functions $C_L(z)$, $C(r,z)$ and $C_P(r,z)$. This allows to check whether the whole cell population is well oxygenated or to evaluate the operating conditions (flow rates, cell density, etc.) required for a suitable oxygenation. Furthermore, the model is very useful in designing the bioreactor, i.e., in calculating number, diameter, and length of fibers. Model equations can be expressed in dimensionless form through the use of the following variables:

$$\bar{C}_L = \frac{C_L}{C_L^0} \quad \bar{C} = \frac{C}{C_L^0} \quad \bar{C}_P = \frac{C_P}{C_P^0} \quad \bar{z} = \frac{z}{L} \quad \bar{r} = \frac{r}{r_e}$$

In this way, model equations become:

$$-\frac{d\bar{C}_L}{d\bar{z}} = \frac{1}{P\acute{e}_{ext}} \left[\bar{C}_L - \bar{C}|_{\bar{r}=1} \right] \tag{5.85}$$

$$\frac{\partial \bar{C}}{\partial \bar{r}} + \bar{r} \frac{\partial^2 \bar{C}}{\partial \bar{r}^2} = \bar{r} \Phi^2 \tag{5.86}$$

$$\left[1 - \bar{r}^2 \left(\frac{r_e}{r_i}\right)^2\right] \cdot \bar{r} \frac{\partial \bar{C}_P}{\partial \bar{z}} = \frac{1}{Pe_{ext}} \left[\frac{\partial \bar{C}_P}{\partial \bar{r}} + \bar{r} \frac{\partial^2 \bar{C}_P}{\partial \bar{r}^2}\right] \tag{5.87}$$

where the following three-dimensionless numbers are introduced:

$$Pe_{ext} = \frac{Q_L}{k_e (2\pi r_e L) n_f} = \frac{Q_L}{k_e S_{ext, fibres}} \tag{5.88}$$

$$\Phi^2 = r_e^2 \frac{q_{max} N}{C_L^0 \pi \left(r_e^2 - r_i^2\right) L n_f \mathscr{D}_{eff}} \tag{5.89}$$

$$Pe_{int} = \frac{2 Q_P}{\mathscr{D}_P \left(\pi \bar{r}_i^2 L\right) n_f} = \frac{u_{max}}{\mathscr{D}_p} \cdot \frac{r_e^2}{L n_f} \tag{5.90}$$

where Φ is the Thiele modulus, Pe_{ext} and Pe_{int} are the modified external and internal Peclet numbers, respectively, $S_{ext, fibres}$ is the area of external surface of fibers, and u_{max} is the maximum velocity at the axis of each fibers in laminar flow.

In terms of dimensionless variables, the boundary conditions are as follows:

$$\bar{z} = 0 \qquad \bar{C}_L (0) = 1 \tag{5.91}$$

$$\bar{C} (\bar{z}, \bar{r}) = \frac{C_P^0}{C_L^0} \bar{C}_P (\bar{r}_i, \bar{z}) \qquad \forall z \tag{5.92}$$

$$\frac{\mathscr{D}_{eff}}{r_e} \cdot \frac{\partial \bar{C} (\bar{z}, \bar{r})}{\partial \bar{r}}\bigg|_{\bar{r}=1} = k_e \left[\bar{C}_L (\bar{z}) - \bar{C} (\bar{z}, \bar{r} = 1)\right] \qquad \forall z \tag{5.93}$$

$$\frac{\partial \bar{C}_P (\bar{z}, \bar{r})}{\partial \bar{r}}\bigg|_{\bar{r}=0} = 0 \qquad \forall z \tag{5.94}$$

$$\mathscr{D}_P C_P^0 \frac{\partial \bar{C}_P (\bar{z}, \bar{r})}{\partial \bar{r}}\bigg|_{\bar{r}=\bar{r}_i} = \mathscr{D}_{eff} C_L^0 \frac{\partial \bar{C} (\bar{z}, \bar{r})}{\partial \bar{r}}\bigg|_{\bar{r}=\bar{r}_i} \qquad \forall z \tag{5.95}$$

$$\bar{C}_P (\bar{z} = 1; \bar{r} = 1) = 1 \tag{5.96}$$

Dimensionless form of model equations allows a reduction of the operating variables required to describe the bioreactor and the more important terms to be easily identified. Therefore, Eqs. 5.85–5.96 are a very useful tool in the analysis of hollow fiber bioreactor behavior.

5.7 Types of Bio-Reactors

Likewise any chemical reactor, from the point of view of mode of operation, bioreactors can be classified as batch, fed batch, or continuous reactors.

In batch bioreactors, no flow of material enters or is drawn out of the reactor. Therefore, a batch bioreactor works at unsteady-state conditions, and mass balances are differential equations.

In fed-batch operation, either an inflow or an outflow of material can be present but not both of them. For instance, after a time of batch operation, an amount of nutrient can be fed to the reactor so that the volume of the reacting mixture increases. A new phase of batch operation follows, at the end of which an amount of the reacting mixture can be discharged and so on. This type of bioreactors is used in cell culture or in enzymatic reactions when there is substrate inhibition. The fed-batch reactor works at unsteady-state conditions or at a *quasi-steady state* if the substrate is consumed as fast as it enters and, in the case of cell cultures, if biomass is not growing.

Continuous bioreactors are open systems where a mass flow rate continuously enters the reactor and exits from it so that steady-state conditions are usually set. Continuous reactors can be well-mixed tanks (CSTR) or tubular reactors where ideal plug flow (PFR) is usually assumed in modeling their operation.

In CSTR, mixing can be supplied by impellers and/or rising of bubble. In the ideal version of CSTR, operating conditions (composition, temperature, etc.) are uniform in the whole volume of the reacting mixture. When suspended cells are cultivated in a CSTR, a recycle from the outflow stream is often used in order to recover cells.

Packed beds bioreactors (PBBs) are the most widely used reactors for cultures of immobilized cells. They consist of a bed of particles containing the cells. The nutrient medium and the fluid to be treated are circulated through the bed. The beads of the packed bed can be non-porous, as glass beads, or porous as ceramic, alginate gel, etc.

In the case of non-porous materials, cells are supported on the external surface of the particles, whereas in porous materials, they are entrapped on the inner pore surface. A list of various materials with high porosity (0.8–0.95) used in biomedical applications is reported in [33]. Using porous particles allows to attain cells densities higher than with non-porous materials. Values considerably higher than densities attainable in suspended cultures and of the order of 10^8 cells/cm^3 have been obtained with porous particles. This value is of the same order of magnitude of the cell density in biological tissues. Therefore, PBBs are currently investigated as promising candidates for bio-artificial organs, in particular for bio-artificial liver (BAL). Beside the cell density, other targets to be attained for developing biomedical devices are as follows:

- A longer viability of the cells, up to 30 days at least for clinical application;
- Adequate oxygenation and other nutrients supply to the cell population;
- Scale up the volume of PBB to about 10^3 cm^3 that is the value required for a clinical application of a BAL.

5.8   An Overview of Bioreactors for Tissue Engineering

Tissue engineering is a new field of biomedical engineering in rapid evolution over the last few years. Its aim is the application of principles and methods of engineering and life sciences to the design, construction, and maintenance of three-dimensional structures able to substitute functions of a living tissue [34]. This target is very ambitious because of the biological complexity of the tissue functions, but its clinical implications are hugely important.

Tissue engineering technologies are based on cell growth, differentiation stimulation of stem cells, and adaptation of the cell population over a synthetic matrix (scaffold) devoted to mimic the role of the extracellular matrix. Cells used in tissue engineering can be differentiated cells (autologous or allogenic), or stem cells able to give rise to different forms of specialized cells under the action of growth factors and other specific chemical and physical regulatory signals.

The first step in generating a 3D tissue structure is cell seeding on the scaffold. This operation is required from the properties of most mammalian cells to be anchorage dependent and from the need of a solid support to reach high densities required for cell–cell communication. An uniform cell seeding is essential for an uniform tissue generation, but, even with small 3D scaffolds, this is still an unsolved technological problem [35].

Scaffold structure and material are of paramount importance in tissue engineering since they must ensure cell attachment, deliver of growth factors and other biochemical factors, diffusion of nutrients, and metabolic products, and finally, they must have specific mechanical properties. For example, they must have a porosity suitable for diffusion of nutrients and other products.

Scaffolds are usually made of biodegradable materials in order to allow their absorption by the surrounding tissues after the surgical installation. Several different materials have been used in scaffold construction. Among them, polylactic acid (PLA), polyglycolic acid (PGA), and polycaprolactone (PCL) are largely used synthetic materials. Further materials tested for scaffold applications are natural materials as collagen, fibrin, chitosan, and glycosaminoglycans (hyaluronic acid).

A number of different techniques, each one with advantages and drawbacks, are used for preparing porous scaffolds, but this issue is not treated in the present chapter.

A key role in tissue engineering is represented by the bioreactor where cell and tissue are cultivated. Mammalian and, especially, human cells are very sensitive to nutrients and toxic by-product concentrations as well as to pH and temperature fluctuations and to mechanical stresses. Therefore, bioreactors must ensure a controlled and stable environment avoiding shear stresses unacceptable for the cell population. Furthermore, bioreactors have to operate under sterile conditions in order to prevent the entry of foreign microorganism.

Due to specificity of tissues and of their cultivation requirements, it is unlike that a single type of bioreactor can be used for all tissue culture operations. However, different specific conditions, such as shape and size, fluid dynamics, and shear stresses, can be imposed on classical types of reactors to meet the requested operation properties. Therefore, stirred (batch or continuous) vessels, packed bed, and fluidized bed reactors and membrane reactors are the typical bioreactors commonly used in tissue engineering. An interesting summary [36] of the characteristics of these bioreactors in tissue engineering applications is reported together with many other comments and a thorough analysis of the problem.

Acronyms

BAL	Bio-artificial liver
CSTR	Continuous stirred batch reactor
OCR	Oxygen consumption rate
PBB	Packed bed bioreactors
PCL	Polycaprolactone
PGA	Polyglycolic acid
PLA	Polylactic acid
PFR	Plug flow reactor

Symbols

C_i	Concentration of the component i
C_P	Product concentration
c_P	Heat capacity
Da	Damköhler number
$\mathscr{D}_e$	Effective diffusion coefficient
$\mathscr{D}_i$	Diffusivity of the component i
E_a	Activation energy
f_A	Fraction of active enzyme
f^+	Fraction of the enzyme with a positive electric charge
f^-	Fraction of the enzyme with a negative electric charge
$f^\pm$	Fraction of the enzyme in the zwitterionic form
J	Diffusive flux
k	Kinetic constant
k_c	Transport coefficient
k^0	Pre-exponential factor in the Arrhenius law
k_M	Michaelis constant in Michaelis–Menten equation
k_S	Constant in Monod equation
k_T	Thermal conductivity
K_d	Equilibrium constant of the enzyme denaturation reaction
K_I	Inhibition constant
K_P	Inhibition constant
n_V	Number of viable cell
n'_V	Number density of viable cell
P	Partition coefficient
Pe	Peclet number
P_m	Membrane permeance
q	Oxygen consumption rate (OCR) per cell
Q	Volumetric flow rate

r	Specific reaction rate (per unit volume)
r_C	Inner core radius
r_{max}	Maximum reaction rate in Michaelis–Menten equation
r^s	Specific reaction rate (per unit surface)
R	Particle radius
R_{diff}	Diffusional resistance
R_{membr}	Membrane resistance
R_{tot}	Total resistance
T	Temperature
$\mathbf{u}$	Velocity (vector)
u_x	Velocity in the direction x
X	Biomass concentration
ΔG^0	Standard change of Gibbs free energy
ΔH	Reaction enthalpy
ΔH^0	Standard change of enthalpy
ΔS^0	Standard change of entropy
η_e	External effectiveness factor
η_T	Overall effectiveness factor
μ	Specific growth rate
μ_{max}	Maximum growth rate
Φ	Thiele modulus
ρ	Density

Subscript

e	External liquid phase
E	Enzyme
E_A	Fraction of the active enzyme
i	Component
in	Internal liquid phase
L	Bulk of liquid phase
P	Plasma
S	Substrate
s	Surface

Superscript

| 0 | Bulk solution |

Part II
Artificial Organs

Chapter 6
Blood Oxygenators and Artificial Lungs

6.1 Introduction

In healthy humans, lungs are the main respiratory organs that provide oxygen to and remove carbon dioxide from the blood. The lungs perform two main functions: ventilation, i.e., the rhythmic movement of air inspiration and expiration, and respiration, i.e., the real gas exchange between air and blood that occurs in the alveoli. Both lung functions may be impaired by many pathologies that modify the capacity of the respiratory system to ensure air flow and gas exchange; when such pathologies become severe, artificial devices to support the lung functions may be required either temporarily or permanently. Furthermore, artificial devices able to replace the lung function are required in cardiac surgery with extracorporeal circulation, when the blood flux is diverted from the heart–lung compartment. Hereafter, the term artificial lung will be used to refer to all devices aimed at providing oxygen and carbon dioxide exchange to replace or to support the function of the natural lungs.

This chapter is aimed at discussing the main engineering aspects involved in the design of artificial lungs, defining the limits of the currently available devices, and understanding the challenges for further developments. To that end, the first part of this chapter provides a short overview of the functions of the respiratory system that allows to define the medical requirements for the artificial devices and the goals that must be achieved in their design; information on the historical development of the artificial lungs is also included. Subsequently, the physical and chemical fundamentals of gas solubility in blood and gas transport through the membranes widely used in artificial lungs are presented. These fundamentals provide the basis for the engineering analysis of the artificial lung and assessment of its performance. This chapter is concluded with a survey of the state of the art of the clinically used devices with indications for possible further improvements.

© Springer-Verlag London 2017
M.C. Annesini et al., *Artificial Organ Engineering*,
DOI 10.1007/978-1-4471-6443-2_6

6.2   Structure and Function of Respiratory System[1]

The respiratory system includes the upper airway (mouth, nose, nasal cavity, pharynx, and larynx), lower airway (trachea, bronchi, bronchioles, and alveoli), and respiratory pump (rib cage, intercostal muscles, diaphragm, and accessory muscles). Bronchioles and alveoli compose the lungs bound by visceral pleura. Visceral pleura reflects in the parietal pleura, on the inner face of the chest, connecting lungs to respiratory pump.

The main function of the respiratory system is connected to the aerobic metabolism of cells, which requires oxygen and produces carbon dioxide. In detail, the respiratory system guarantees the elimination of the CO_2 and the uptake of the O_2 in two phases: ventilation and respiration. Ventilation involves the respiratory pump and airway conducts, till bronchioles, while respiration involves alveoli and blood, through the respiratory membrane (external respiration) as well as blood and tissue through the capillary membrane (internal respiration).

Ventilation consists of a mechanic and rhythmic process in which air movement in (inspiration) and out (expiration) of lungs is permitted. Air flux is due to a transpleural pressure gradient, which determines convection movements. At rest, during inspiration, the contraction of diaphragm and intercostal muscles leads to the expansion of chest cavity and lungs, which, due to a reduction of intrapleural pressure, determines air inflow. During expiration, the relaxation of muscles leads to elastic return of the chest and lungs, with an increase in intrapleural pressure, which causes the outflow. Through pressure variation at each respiratory act, the movement of 500 ml of air is realized (tidal volume). With respect to this volume, only 350 ml participates in the real gas exchange: in fact, 150 ml remain in the conducting airway (anatomic dead space[2]) that does not take part in gas exchange. The rhythmic function of ventilation has a rate of about 15 acts/min, determining a volume rate of about 7.5 l/min (total ventilation = tidal volume × respiratory rate). Considering the dead space, the alveolar ventilation (the total volume of air arriving to alveoli in a minute) is about 5.25 l/min.

Respiration is realized in the alveoli that are the real functional parts of the whole airway. Alveoli have a mean diameter of 200–300 μm and represent the distal part of the lungs in which air is conducted by a terminal airway ramification (terminal bronchioles). It has been estimated that in a mean adult, the respiratory zone of the lungs has a surface of about 100–150 m^2 and contains a volume of 2.5–3 l of air. The effectiveness of the large surface of the respiratory zone in providing gas exchange is further increased by the characteristics of the alveolar capillaries and alveolar membrane (blood–gas interface). Alveolar capillaries are large enough to permit the passage of a single red cell (7–10 μm), leading to a continuous sheet of blood flowing

---

[1] This and the following section were authored by Felice Eugenio Agrò, Marialuisa Vennari, and Maria Benedetto—University School of Medicine "Campus Bio-Medico" of Rome, Italy.

[2] A further dead space (physiologic dead space) is considered as a functional evaluation of that part of lung which does not eliminate CO_2. The anatomic and physiologic dead spaces are normally very similar, but they may differ in presence of many lung diseases.

over the alveolar membrane and allowing a sufficient contact for gas exchange even in a short time (0.75 s per red cell, 0.25 on exertion). Moreover, the alveolar membrane is extremely thin ($1.5\,\mu m$), enhancing the gas exchange rate between the alveolar gas and blood. As previously reported, both O_2 and CO_2 must be exchanged between the alveolar gas and blood to balance the gas consumption and production due to the cell aerobic metabolism. Actually, O_2 diffuses from alveoli (O_2 pressure of 150 mmHg) to venous blood (O_2 pressure of about 40 mmHg), while CO_2 diffuses from venous blood (CO_2 pressure about 45 mmHg) to alveoli (CO_2 pressure of 40 mmHg). It is worth noting that since the ratio of CO_2 production to O_2 consumption by tissues (respiratory exchange ratio) is about 0.8, an adequate respiratory function requires the same ratio to apply to the rate of CO_2 elimination and O_2 uptake in lungs.

As discussed later in detail (Sect. 6.7), gas exchange efficiency depends on ventilation and perfusion. Different parts of the lungs are not equally perfused nor ventilated; in particular, the cardiac output is distributed in the different parts of the lungs according to the transmural vessel pressure (the difference between the capillary and alveolar pressures). Actually, in the upper part of the lung (apical lung), the pulmonary arterial pressure (i.e., the pressure of the blood entering the pulmonary capillary) is lower than the alveolar pressure, this tends to narrow the cross section of capillaries, and only minimal blood flux is permitted; as a consequence, this region is only minimally participating to gas exchange. In the middle and basal lung, arterial pulmonary pressure is higher than alveolar pressure and a higher blood flux is permitted: in the middle zone, the arterial pressure is higher, but the venous pressure (i.e., the pressure of the blood at the end of the pulmonary capillary) is lower than alveolar pressure and the blood flux is determined by arterial–alveolar pressure difference; on the other hand, in the basal zone, both the arterial and venous pulmonary pressures are higher than alveolar pressure and the blood flux is determined by the arterial–venous pressure difference. The middle zone is the ideal zone where ventilation/perfusion rate is near to 1. Actually, while arterial oxygen concentration is largely affected by the ventilation–perfusion ratio, the CO_2 concentration is mainly dependent on ventilation. As a consequence, an increase in alveolar ventilation may correct CO_2 elimination, while it does not surely correct hypoxemia caused by alterations of the ventilation–perfusion ratio.

6.3 Extracorporeal Gas Exchange Devices in Clinical Practice

Extracorporeal devices able to provide gas exchange are routinely used intraoperatively in cardiac surgery with extracorporeal circulation. In this case, usually a membrane oxygenator is included as an essential part of the cardiopulmonary pump needed to replace the cardiac and pulmonary function; cardiopulmonary bypass (CPB) is generally applied for few hours, but, in some cases, a prolonged support may be needed. In these cases, extracorporeal membrane oxygenation (ECMO) is used as life support or lung assist, to provide continuous support, typically for a period of the order of days to weeks.

The use of ECMO to supplement the insufficiencies or the failure of the respiratory system is less frequent, but the interest in this procedure has been increasing in recent years. In fact, in clinical practice, the two functions of the respiratory system (ventilation and respiration) may be affected by many pathologies: both primitive lung diseases—e.g., pneumonia, chronic obstructive pulmonary disease (COPD), fibrosis, and acute respiratory distress syndrome (ARDS)—and secondary pulmonary involvements—e.g., cardiogenic edema, neurological impairment, and chest alterations—may modify the capacity of the respiratory system to ensure sufficient air flux and gas exchange. When the decrease of this capacity is life-threatening, the functions of the respiratory system should be artificially supported. In this case, mechanical ventilation (MV) and/or extracorporeal oxygenators (EOs) may be indicated. Historically, ECMO has been used with benefits in neonates and children with reversible cardiorespiratory failure. In recent years, evidence supporting ECMO use in adults has emerged, with increased survival rate compared to the optimization of standard therapy both in severe respiratory failure and cardiac failure. This evidence has been underlined during H1-N1 epidemic, in CESAR [37] trials on patients with ARDS, and in reports on patients with cardiac arrest, cardiogenic shock, and who failed weaning from CPB (Extracorporeal Life Support Organization, ELSO, registry).

Hereafter, the focus will be put on ECMOs. Indeed, ECMO seems to present some advantages compared to standard therapy based on mechanical ventilation: in particular, the ventilator-induced lung injury (barotrauma or volutrauma) is avoided, and the lungs are allowed to rest and acute damage recovery.

Two main types of ECMO may be distinguished: veno-venous (VV) and veno-arterial (VA).[3] VA ECMO supports both pulmonary and cardiac functions, while VV ECMO provides only respiratory support. The main indications to ECMO use are acute cardiac failure (AV ECMO) and acute respiratory failure (VV ECMO) with high mortality risk, despite optimal conventional therapy. In particular, according to ELSO guidelines for ECMO centers, the main indications for ECMO use in adults are as follows:

- acute respiratory failure with a ratio of arterial oxygen pressure[4] (mmHg) to the fraction of oxygen in the inspired air (PaO2/FiO2) < 150 on FiO2 > 90 % and/or Murray score[5] 2–3 (ECMO suggested) or with a PaO2/FiO2 < 80 on FiO2 > 90 % and Murray score 3–4 (ECMO indicated);

---

[3]Pumpless arterio-venous ECMO, using the patient's own arterial pressure to pump blood through the circuit, may also be used, mainly for carbon dioxide removal [38].

[4]In the medical field, this parameter is usually indicated with the symbol PaO2. However, in the equations reported later in this chapter, the symbol p_{a,O_2} will be rather used in order to be consistent with the overall notation.

[5]The Murray score evaluates lung pathology severity according to PaO2/FiO2, positive end-expiratory pressure (PEEP) value, chest radiography, and lung compliance.

- CO_2 retention with PaCO2 > 80 mmHg or inability to achieve safe inflation pressures;
- severe air leak syndromes (pneumothorax, broncho-pleural fistula);
- refractory cardiogenic shock;
- septic shock with severe cardiac dysfunction (indication is some center);
- failure to wean from cardiopulmonary bypass;
- cardiac arrest;
- as a bridge to the placement of a ventricular assist device (VAD), to cardiac transplantation, and during the recovery of revascularization in myocardial infarction, myocarditis, and postcardiotomy. In case of respiratory failure, generally, there are no absolute contraindications to ECMO use. Relative contraindications are conditions with known poor outcome despite ECMO (MV > 7 days with high pressure) and specific patient conditions (e.g., severe obesity and comorbidities).

Recently, ECMO use has been suggested in case of chronic pulmonary diseases such as COPD. The use of systems for specific CO_2 elimination may reduce the incidence of intubation in case of COPD exacerbation in association with noninvasive ventilation (NIV). These results may be encouraging to further experience ECMO use in chronic respiratory failure.

In pediatric cases, ECMO is indicated within the first week of MV at high pressure or when a shock refractory to standard treatment developed, the weaning from CPB failed, after a successful cardiopulmonary reanimation the patient is still unstable, or a severe cardiac failure of any etiology develops. Contraindications depend on age, comorbidities, and the presence of contraindication to anticoagulation.

6.4 General Remarks on Blood Oxygenator Design

Before going into details on the study of blood oxygenators, it is beneficial to briefly discuss the requirements for these devices and the goals to achieve in their design. Thus, it will be easier to understand the technological development of blood oxygenators, the present-day applications, and the work that is still to be done to obtain better devices for wider application fields.

A device capable of providing the complete respiratory function, as in a cardiopulmonary bypass for an open-heart surgery, will be considered: such device must transfer up to 250 Nml/min (11 mmol/min) of oxygen, in order to meet the basal metabolic requirement of an adult patient; in the past, the gas exchange requirement was reduced by lowering the patient's body temperature, but nowadays, there is a trend toward normothermic perfusion. At the same time, the device has to remove about 200 Nml/min (9 mmol/min) of carbon dioxide. The CO_2-to-O_2 rate ratio must equal the respiratory ratio (0.8), and a fine control of the amount of carbon dioxide removed is required to avoid both hyper- or hypocapnia.

Typical operating conditions are summarized in Fig. 6.1. The whole blood flow rate (5 l/min) is pumped through the device, with a roller or centrifugal pump; the

Fig. 6.1 Scheme of
operating conditions for an
oxygenator used in
cardiopulmonary bypass

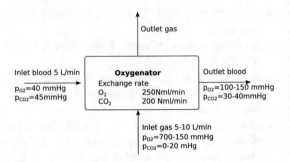

inlet blood has O_2 and CO_2 partial pressures of about 40 mmHg (corresponding to
about 6.6 mM, see Sect. 6.6.1.1) and 45 mmHg (corresponding to about 12.7 mM, see
Sect. 6.6.1.2), respectively; the arterialized blood has to attain O_2 and CO_2 partial
pressures of about 100–150 mmHg (8.7–8.9 mM) and 30–40 mmHg (8.5–11 mM),
respectively.

The inlet gas is usually pure oxygen or an oxygen-rich mixture, which provides
a large driving force for oxygen transfer to the blood. On the other hand, the gas
flow rate controls the carbon dioxide concentration in the sweep gas and, therefore,
the effectiveness of its removal from blood: the higher the gas flow rate, the lower
of the carbon dioxide pressure in the sweep gas and the higher the driving force for
carbon dioxide removal [39]. Therefore, the carbon dioxide transfer rate depends
directly on the gas flow rate. More specifically, since CO_2 partial pressure in the
outlet gas cannot exceed 40–45 mmHg in order to ensure a sufficient driving force
for its removal, a minimum gas flow rate of about 3.6 l/min is required.[6] Usually, the
actual gas flow rate ranges from 5 to 10 l/min.

Regardless of the type of device used, the gas transfer rate is given by:

$$\text{exchange surface} \times \text{mass transfer coefficient} \times \text{driving force}$$

or

$$\text{volume} \times \frac{\text{exchange surface}}{\text{volume}} \times \text{mass transfer coefficient} \times \text{driving force}$$

From a clinical point of view, in order to minimize the transfusion of donor blood
or plasma expander solution, it is desirable to minimize the oxygenator priming volume as well as the volume of the extracorporeal circuit; therefore, it is of paramount
importance that the oxygenator be designed so as to have a high exchange surface
area to volume ratio. On the other hand, the exchange surface area should be kept
as low as possible, still meeting the gas transfer rate requirements. Indeed, blood
exposure to exogenous surfaces causes the activation of the complement and coag-

---

[6]This estimation is based on a CO_2 flow rate removed of 0.2 Nl/min and outlet partial pressure of
40 mmHg: $0.2 \, Nl/min \cdot 760/40 = 3.6 \, Nl/min$.

ulation cascade, which impair the biocompatibility of the device; furthermore, for membrane oxygenators, a reduction of the membrane area will also reduce the cost of the device. Therefore, a fundamental target to be pursued in the design of blood oxygenators is to obtain high mass transfer coefficients and exploit the maximum driving force available.

As for the mass transfer coefficient, this parameter is mainly determined by the blood-side gas transport resistance and, in membrane oxygenators, also by the mass transfer resistance of the membrane itself; therefore, the new technological developments should be aimed both at optimizing blood fluid dynamics in the oxygenators, in order to improve gas transport in the blood layer, and producing membranes with high gas permeability.

The maximum available driving force for gas transfer is about 650 mmHg for oxygen[7] and 40 mmHg for carbon dioxide. In order to mimic the respiratory exchange ratio and best exploit the maximum available driving force for both gases, the CO_2-to-O_2 mass transfer coefficient ratio should approach $0.8 \times 650/40 \simeq 13$. For lower mass transfer coefficient ratios, carbon dioxide removal controls the exchange surface required; on the other hand, for higher mass transfer coefficient ratios, the CO_2 removal rate must be controlled by proper choice of the gas flow rate and/or carbon dioxide concentration in the inlet gas.

6.5 Development of Blood Oxygenators: History and Current Solutions

The birth of blood oxygenators is strictly related to the requirements set by the cardiopulmonary bypass (CPB) procedure in open-heart surgery; therefore, the early devices developed were integrated systems including both a blood pump and an oxygenator.

The first devices used between 1930 and 1960 in open-heart surgery provided O_2 and CO_2 exchange by direct contact of blood with a gas phase. Two methods were used to obtain a large gas–blood contact area: to contact the gas phase with a thin blood film flowing on a solid surface (film oxygenators) or to disperse small gas bubbles into venous blood (bubble oxygenators) [40].

In film oxygenators, a thin blood film is formed on stationary or rotating surfaces; gas exchange occurs through the film surface directly exposed to a gas phase with high oxygen and low carbon dioxide partial pressure. In the first successful CPB operation, a screen oxygenator was used. Such device consisted of a series of upright wire mesh screens with the venous blood flowing from the top by gravity and forming a thin film on the screens; arterialized blood was collected at the bottom of the screens to be returned to the patient. Screen oxygenators were included in a commercial device for

---

[7] If pure oxygen is used, about 50 mmHg of water vapor is required to properly humidify the inlet gas.

Fig. 6.2 Scheme of a
rotating disk oxygenator

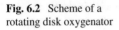
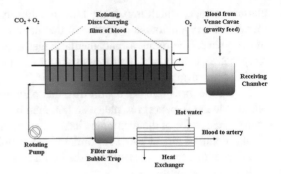

CPB (Mayo-Gibbon pump oxygenator), but such apparatus was bulky and required a large blood and saline solution priming volume.

In the mid-fifties, rotating disk oxygenators were also introduced in the clinical practice. In this type of oxygenators (see Fig. 6.2), a blood film is created on vertical disks rotating on an horizontal axis and dipping into a pool of venous blood; in the upper part of the disks, the blood film is exposed to an oxygen-rich atmosphere, thus resulting in a good oxygenation capacity. In spite of the large priming volume required and assembling and sterilization issues, rotating disk oxygenators were largely used in the clinical practice until the seventies: this was mainly due to the perception that blood trauma caused by this type of device is generally quite low [40].

Almost in the same period (1950s), the idea to obtain a large gas–blood contact area and efficient gas transfer by bubbling an oxygen-rich gas phase into venous blood was explored. In spite of the apparent simplicity of this approach, which also allows to operate with a low priming volume, several issues had to be solved before a bubble oxygenator suitable for clinical application could be obtained. First of all, gas bubble dispersion in blood required a defoaming system to remove gaseous emboli before returning the arterialized blood to the patient; in this respect, the size of gas bubbles is crucial: while smaller gas bubbles, with high surface-to-volume ratio, result in a larger specific surface area and more efficient gas exchange, they are less prone to rise spontaneously to the surface and are more difficult to remove. Furthermore, the need for an adequate ratio between the oxygen transfer flow rate (which is kinetically limited by the exchange surface area) and carbon dioxide transfer flow rate (which is limited by CO_2 accumulation in the gas phase) had to be compromised to achieve an efficient device.

In 1955, a helical reservoir pump oxygenator (De Wall oxygenator [41]) was used for the first time in an intracardiac surgical operation: in such oxygenator (see Fig. 6.3), oxygen is mixed with venous blood in a vertical cylinder where gas exchange occurs; on top of the mixing tube, the two fluid phases enter a silicon-coated chamber, where gas bubbles coalesce and some debubbling occurs; finally, debubbling is pushed further in a helical tubular reservoir, in which the gas bubbles float upward while blood flows downwards. De Wall bubble oxygenator gained a large acceptance, being used in 90 % of open-heart operations in 1976 [40]; indeed,

Fig. 6.3 Scheme of the De Wall bubble oxygenator

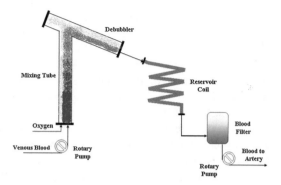

such device was cost-effective and easy to assemble and sterilize. The popularity of this device was also supported by the introduction of a single-use, presterilized, and prepacked plastic version [41], together with different systems to improve defoaming (see for example [42]). Even if a high oxygenation efficiency can be achieved with bubble oxygenators, two important shortcomings limit their use: firstly, the need for a thorough removal of gas bubbles in order to reduce the risk of embolism; secondly, the blood damage due to mechanical stress and the direct contact with gas and solid surfaces. In particular, blood damage issues limit the duration of sessions with direct contact oxygenators and made them unsuitable for long-term therapeutic support, as required for patients with ARDS or newborns with neonatal respiratory distress syndrome.

A significant improvement to the state of the art was obtained with the introduction of membrane oxygenators, in which a gas permeable membrane is interposed between the gas phase and blood, thus avoiding the need of a defoaming unit and significantly reducing blood damage. Indeed, membrane oxygenators mimic the natural lung configuration, where gas exchange occurs through the alveolar membrane. Actually, the first idea to use membranes for blood oxygenation dates back to the mid-forties, when Kolff noted that during hemodialysis sessions, blood oxygenation occurred in parallel with detoxification, due to gas exchange with the oxygen-saturated dialysis solution [43]; nevertheless, more than twenty years were necessary to develop effective membrane oxygenators for clinical use and to challenge the dominant position held by bubble oxygenators [44]. In the early developments, the focus was put on producing reliable membranes with high permeability to both O_2 and CO_2. Polyethylene membranes showed a low CO_2 permeability (only five times greater than oxygen permeability, compared to a permeability ratio in the lung membrane exceeding 20): as a result, the membrane surface area was controlled by the required CO_2 transfer rate. A significantly better performance was obtained in the sixties with silicone rubber membranes, which showed higher permeabilities and a more favorable CO_2 to O_2 transfer ratio; on the other hand, with the improvement of membrane performance, oxygen diffusion through the blood film became a significant resistance to the gas transfer.

The development of membrane oxygenators is then the result of advancements on membrane materials, leading to improved permeability and selectivity, and optimization of fluid dynamics of the devices, which allows to reduce both the membrane surface area and the priming volume. Spiral coil (Kolobow oxygenators) as well as plate and screen configurations were firstly used; the hollow fiber configuration, already adopted in hemodialysis, was transferred to blood oxygenators in the middle of seventies [45–47].

The next major advance came with the introduction of microporous hydrophobic membranes with a pore size below 1 μm: in these membranes, O_2 and CO_2 diffusion does not occur through the membrane material but in the gas-filled pores. The use of highly hydrophobic materials prevents plasma leakage through the membrane pores: at the beginning of eighties, the first commercial hollow fiber oxygenator used silicone-coated microporous polypropylene membranes; more recently, poly(4-methyl-1-pentene) (commonly called with the trademark TPX) membranes showed a very good performance [48]. Although microporous membranes exhibit high gas transfer rates, in long-term use they undergo a progressive alteration of surface properties due to protein and lipid adsorption, the wetting of membrane pores, and the plasma infiltration and leakage through the pores; as a consequence, the membrane permeance markedly reduces over time. Composite hollow fiber membranes, including a non-porous polymer layer ("skin") on the microporous membrane surface, that prevent the blood infiltration into the pores and control the long-term performance have been developed in the last years.

Blood oxygenators currently used in clinical practice are membrane oxygenators, in most cases based on microporous hollow fiber membranes. The fibers (diameter 200–300 μm) are wound or bundled in a hard shell, with a fiber packing density of 40–60 % and a large surface area-to-volume ratio (1–2.5 m², with a priming volume in the range of 100–350 ml). The gas usually flows inside the lumen of the fibers, while blood flows outside, through the fiber bundle (*extraluminal flow*). The opposite flow pattern, with blood flowing inside the fibers (*intraluminal flow*), is also possible, but less frequent; indeed, extraluminal flow results in a lower resistance to gas transfer through the blood film, because the flow past the fibers induces a secondary flow that enhances mixing; furthermore, since the extraluminal side has a wider cross section, also the resistance to blood flow and pressure drop are reduced in extraluminal flow. A passive secondary flow may be also obtained by putting obstacles in the blood path, creating undulation or texturing the membrane surface or using a specific flow geometry (e.g., helical coil). It is important to note that turbulence and secondary flow increase the pressure drop and the shear rate experienced by red cells; therefore, also the shear-induced hemolytic damage is increased.

In extraluminal flow, the angle between the fibers and blood flow affects the mass transfer coefficient [49]. Good results have been obtained with blood flowing perpendicularly to the hollow fiber axis, but different flow patterns are found in the different commercial devices: blood flows radially through the fiber bundle in the Medtronic Affinity® NT oxygenator (Fig. 6.4) or perpendicularly to the gas pathway in the Maquet Quadrox® (Fig. 6.5). Silicone oxygenators with non-porous silicone sheets have poorer performance in terms of gas exchange efficiency, but are clinically

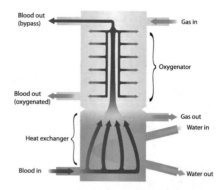

Fig. 6.4 Flow pattern in Affinity® NT blood oxygenator, including a heat exchanger

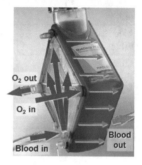

Fig. 6.5 Picture of the Maquet Quadrox®. Flow pattern highlighted. Reproduced from [50], with permission

used for long-term support. Many of the modern devices are integrated with a rigid reservoir and a heat exchanger [45].

Configurations and performance of some membrane oxygenators used in the clinical practice are summarized in Table 6.1.

6.6 Fundamentals of Gas Exchange

The knowledge and quantitative description of gas solubility and transport in blood are required in order to understand gas exchange both in healthy lungs and in blood oxygenators. This section offers a brief overview of these fundamental aspects with specific focus on respiratory gases, i.e., O_2 and CO_2.

6.6.1 Gas Solubility in Blood

The dissolution of both O_2 and CO_2 in blood follows a combined physical and chemical mechanism: both gases are absorbed into plasma as molecular species and then take part in chemical reactions in the liquid phase, which enhance their solubility.

Table 6.1 Properties of hollow fiber membrane blood oxygenators

	Membrane		Priming volume ml	Blood flow l/min	Gas transfer rate	
	Material	Area, m^2			O_2, ml/min	CO_2, ml/min
Terumo, Capiox® FX	PP	0.5–2.5	43–260	0.1–7	50–500	50–500
Medtronic Affinity® NT	PP	2.5	270	1–7	50–400	50–400
Maquet Quadrox®	PP	1.8	250	0.5–7	max 425	max 320
Medos hilite®	PP	0.39–1.9	57–275	1–7		
Medos hilite® LT	PMP	0.32–1.9	55–275	0.8–7	100–550	75–350
GISH Biomedical Vision®	PP	2.45	280	1–8	400[a]	200–500[a]
NIPRO Vital	PP	2	180	0.5–7		

PP: polypropylene; PMP: polymethylpentene
[a] Blood flow rate 6 l/min

More specifically, molecular O_2 binds to hemoglobin, while CO_2 dissociates leading to the formation of carbonate and bicarbonate ions[8]; furthermore, CO_2 and O_2 chemical equilibria are coupled via the pH effect with oxyhemoglobin dissociation and CO_2 binding to hemoglobin.

Since the early works of Adair [51], several studies have been reported in the literature and extensive reviews have also been published [52, 53]; here, the main results and theoretical models useful for the design of blood oxygenators are reported.

6.6.1.1 Oxygen Solubility

Molecular oxygen (physical) solubility in plasma is described by Henry's law[9]

$$p_{O_2} = H'_{O_2} c_{O_2,m} \tag{6.1}$$

---

[8] A small fraction of carbon dioxide (about 5 %) is also bound to hemoglobin.

[9] Sometimes, in biomedical literature, the solubility of molecular oxygen in a liquid phase is written as

$$c_{O_2,m} = \alpha p_{O_2}$$

Obviously, $\alpha = 1/H'$.

where p_{O_2} is the oxygen partial pressure in the gas phase, $c_{O_2,m}$ is the concentration of free, molecular oxygen in the liquid phase, and H'_{O_2} is the Henry constant of molecular oxygen in plasma. The value of H'_{O_2} ($1 \cdot 10^6$ atm mol^{-1}cm^3) denotes a very low solubility of oxygen in plasma: indeed, the physical solubility alone is insufficient to ensure sufficient oxygen transport to fulfill the tissue metabolic requirements, which explains why the presence of an oxygen carrier in blood is necessary.[10]

Oxygen transport is facilitated by its binding to hemoglobin inside red blood cells; in this tetrameric protein, each amino acid chain contains a heme group which binds to an oxygen molecule and a terminal amino group which can bind to a CO_2 molecule. As for oxygen–hemoglobin binding, Adair [51] suggested to account for four binding reactions between the partially oxygenated hemoglobin, indicated as $Hb(O_2)_{n-1}$, and the oxygen molecule. The general form of such reactions is as follows:

$$Hb(O_2)_{n-1} + O_2 \rightleftarrows Hb(O_2)_n \qquad n = 1, \ldots, 4$$

and the corresponding equilibrium constants are as follows:

$$K_n = \frac{c_{Hb(O_2)_n}}{c_{Hb(O_2)_{n-1}} p_{O_2}} \tag{6.2}$$

Accounting for these reactions, a fractional saturation of hemoglobin, defined as

$$S_\% = \frac{\text{oxygen molecules bound to hemoglobin}}{\text{maximum oxygen molecules bound to hemoglobin}}$$

can be evaluated as:

$$
\begin{aligned}
S_\% &= \frac{\sum_{n=1}^{4} c_{Hb(O_2)_n}}{4 c_{Hb}} \\
&= \frac{K_1 p_{O_2} + 2K_1 K_2 p_{O_2}^2 + 3K_1 K_2 K_3 p_{O_2}^3 + 4K_1 K_2 K_3 K_4 p_{O_2}^4}{4\left(1 + K_1 p_{O_2} + K_1 K_2 p_{O_2}^2 + K_1 K_2 K_3 p_{O_2}^3 + K_1 K_2 K_3 K_4 p_{O_2}^4\right)}
\end{aligned} \tag{6.3}
$$

where c_{Hb} represents the hemoglobin concentration[11]; typical values of the binding equilibrium constants for human blood are reported in Table 6.2: the cooperativity of heme groups for O_2 binding, i.e., the increase hemoglobin affinity for oxygen after binding the first O_2 molecule, results in a sigmoid saturation curve; as for many engineering calculations, such a curve can be described by the simplified Hill equation [55]:

---

[10]With an oxygen pressure of about 160 mmHg in the gas, molecular oxygen concentration in blood at equilibrium is about 0.21 mM. With such a concentration, a blood flow of 5 l/min carries only 1 mmole/min of molecular oxygen, i.e., about 10 % of the metabolic requirement.

[11]Indeed, hemoglobin is enclosed in red blood cells, but here, we refer to the average hemoglobin concentration in blood. For healthy humans, a mean value of $c_{Hb} = 2.2$ mM can be considered.

Table 6.2 Equilibrium constant for oxygen-hemoglobin-binding reaction ($mmHg^{-1}$) [54]

K_1	K_2	K_3	K_4
0.004 ± 10 (%)	0.043 ± 18 (%)	0.262 ± 58 (%)	0.039 ± 52 (%)

Table 6.3 Dependence of p_{50} on several physiological variables (p_{50}, p_{CO_2}, mmHg; T,K) [56, 57]

- Dependence on diphosphoglycerate-to-hemoglobin ratio and CO_2 partial pressure

$$\log p_{50} = f(p_{CO_2})\log \frac{c_{DPG}}{c_{Hb}} + g(p_{CO_2})$$

$$f(p_{CO_2}) = A \cdot p_{CO_2} + B$$

$$g(p_{CO_2}) = C \cdot p_{CO_2} + D$$

constants A, B, C, and D depend on pH

	pH = 7	pH = 7.6
A	$-0.69117 \cdot 10^{-3}$	$-0.1380 \cdot 10^{-2}$
B	0.3365	0.3607
C	$0.3598 \cdot 10^{-3}$	$0.9089 \cdot 10^{-3}$
D	1.599	1.360

- Dependence on temperature

$$\frac{d \log p_{50}}{d(1/T)} \approx -2150$$

$$S_{\%} = \frac{\left(p_{O_2}/p_{50}\right)^n}{1 + \left(p_{O_2}/p_{50}\right)^n} \qquad (6.4)$$

which is based of the following apparent reaction

$$Hb + nO_2 \rightleftarrows Hb(O_2)_n \qquad (6.5)$$

In Eq. 6.4, p_{50} is the oxygen partial pressure at which 50 % of hemoglobin oxygen-binding sites are saturated (the higher the p_{50} value, the lower the hemoglobin affinity for oxygen), while n is defined by Eq. 6.5. A value for n of 2.7 was found to fit well data for normal human blood in the saturation range of 20–98 %, while a value of about 27 mmHg is usually assumed for p_{50}; however, it is well known that p_{50} is affected by temperature, CO_2 concentration, pH, and ratio of diphosphoglycerate[12] (DPG) to hemoglobin concentrations. More specifically, p_{50} is a decreasing function of temperature, CO_2 and DPG concentration, and an increasing function of pH. Several correlations have been proposed to describe the dependence of p_{50} on the above-listed variables; as an example, those reported by Samaja et al. [56, 57] are reported in Table 6.3.

---

[12]Diphosphoglycerate is a product of hemoglobin metabolism.

Fig. 6.6 Oxygen
concentration in blood as a
function of O_2 partial
pressure

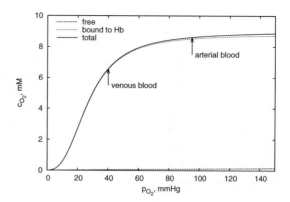

Finally, the total oxygen concentration in blood is given by:

$$c_{O_2} = \frac{p_{O_2}}{H'_{O_2}} + 4 \frac{\left(p_{O_2}/p_{50}\right)^n}{1 + \left(p_{O_2}/p_{50}\right)^n} c_{Hb} \tag{6.6}$$

Figure 6.6 reports the total oxygen concentration in blood as a function of oxygen partial pressure in the gas phase: it can be seen that more than 99 % of the dissolved oxygen is present as oxyhemoglobin complex; in other words, the capacity of blood to transport oxygen to peripheral tissues is strongly affected by its hemoglobin content. In Fig. 6.6, the regions of the plot corresponding to oxygen pressures in venous and arterial blood are marked: it is evident that the effect of a change in p_{O_2} on oxygen solubility is much stronger in venous rather than in arterial blood. Furthermore, it is worth noting that the p_{O_2} range from 30 to 50 mmHg, which corresponds to the oxygen level that must be maintained in the peripheral tissues, is also characterized by a strong dependence of oxygen solubility on p_{O_2}: this feature allows to have a significant O_2 exchange rate between blood and tissues even with small p_{O_2} differences.

By differentiating Eq. (6.6), the following expression is obtained:

$$\frac{\partial c_{O_2}}{\partial p_{O_2}} = \frac{1}{H'_{O_2}} + 4 c_{Hb} \frac{\partial S_\%}{\partial p_{O_2}} = \frac{1}{H'_{O_2}} \left(1 + \mathcal{K} c_{Hb}\right) \tag{6.7}$$

where $\mathcal{K} = 4 H'_{O_2} \left(\partial S_\%/\partial p_{O_2}\right)$. In the physiological blood oxygen pressure range (40–95 mmHg), a linearized form of Eq. 6.6 with a constant value of about $12.1 \cdot 10^6$ mol^{-1}cm^3 for $\mathcal{K}$ provides a reasonable approximation of the oxygen solubility curve; therefore, in this partial pressure range, an *apparent Henry's constant* for O_2 in blood can be used to describe O_2 solubility

$$\mathcal{H}_{O_2} = \frac{\Delta p_{O_2}}{\Delta c_{O_2}} = \frac{H'_{O_2}}{\left(1 + \mathcal{K} c_{Hb}\right)} \simeq 27.8 \text{ mmHg mmol}^{-1} \text{l} \tag{6.8}$$

Table 6.4 CO_2 content in arterial and venous blood (concentrations are referred to the blood volume; Htc $= 45\,\%$) [58]

p_{CO_2}, mmHg	Arterial	Venous		
	40	46		
	Plasma		Red cells	
	Arterial	Venous	Arterial	Venous
pH	7.4	7.37	7.2	7.175
CO_2, mM	0.68	0.78	0.4	0.46
HCO_3^-, mM	13.52	14.51	5.01	5.46
Carbamino[a], mM	0.30	0.30	0.75	0.84

[a]Bound to hemoglobin (in red cells) or plasma proteins

6.6.1.2 Carbon Dioxide Solubility

While almost all oxygen dissolved in blood is bound to hemoglobin inside red cells, the majority of carbon dioxide is found in plasma (2/3) and red cells (1/3) as bicarbonate ion and only less than 5 % is bound to hemoglobin. Table 6.4 reports some typical values of the carbon dioxide concentration in arterial and venous blood. Distribution of CO_2 among different chemical species, namely molecular carbon dioxide, bicarbonate, and carbonate ions, occurs according to the following hydrolysis reactions:

$$CO_2 + H_2O \rightleftarrows H_2CO_3 \quad K_h = 1.7 \cdot 10^{-3}$$

$$H_2CO_3 \rightleftarrows HCO_3^- + H^+ \quad K_{a1} = 2.5 \cdot 10^{-4}\,M$$

$$HCO_3^- \rightleftarrows CO_3^{2-} + H^+ \quad K_{a2} = 5.6 \cdot 10^{-11}\,M$$

By accounting for the above-listed reactions, the total carbon dioxide concentration can be expressed as:

$$c_{CO_2} = c_{CO_2,m} + c_{HCO_3^-} + c_{CO_3^{2-}} = c_{CO_2,m} \left(1 + \frac{K_h K_{a1}}{10^{-pH}} + \frac{K_h K_{a1} K_{a2}}{10^{-2\,pH}}\right) \quad (6.9)$$

In Eq. 6.9, CO_2 in the form of carbonic acid or carbamino compounds (bound to hemoglobin) was neglected. At the physiological blood pH (about 7.3), the ratio of HCO_3^- to CO_2 concentrations is about 93 %.

Assuming that free CO_2 solubility is described by the Henry's law, the total CO_2 concentration in the liquid phase can be expressed as a function of its partial pressure in the gas as

$$c_{CO_2} = \frac{p_{CO_2}}{H'_{CO_2}} \left(1 + \frac{K_h K_{a1}}{10^{-pH}} + \frac{K_h K_{a1} K_{a2}}{10^{-2pH}}\right) \quad (6.10)$$

Therefore, the solubility of CO_2 in blood can be expressed by referring to a pH-dependent apparent Henry's constant, $\mathscr{H}_{CO_2}$, which can be calculated as follows

$$\frac{1}{\mathscr{H}_{CO_2}} = \frac{c_{CO_2}}{p_{CO_2}} = \frac{1}{H'_{CO_2}}\left(1 + \frac{K_h K_{a1}}{10^{-pH}} + \frac{K_h K_{a1} K_{a2}}{10^{-2pH}}\right) \qquad (6.11)$$

It is worth noting that blood pH depends on CO_2 concentration and venous blood has a slightly but significantly lower pH than arterial blood. As a consequence, a nonlinear relation between c_{CO_2} and p_{CO_2} is actually observed. However, as previously shown for O_2 solubility (see Eq. 6.8), in the relevant range of physiological conditions for gas exchange, the CO_2 solubility curve can be linearized and an approximately constant value for $\mathscr{H}_{CO_2}$ assumed, that is:

$$\mathscr{H}_{CO_2} = \frac{\Delta p_{CO_2}}{\Delta c_{CO_2}} = 5 \cdot 10^{-3} \text{ mmHg mol}^{-1} \text{ l} \qquad (6.12)$$

6.6.2 Gas Transport in Blood

The analysis of gas solubility presented in Sect. 6.6.1 showed that dissolved O_2 and CO_2 are present in blood both as free molecular and chemically bound species. This aspect affects also the transport of gases in blood, especially with regard to diffusive transport.

This section deals with the analysis of gas transport in blood. The contents of the first part of this section apply equally to oxygen and carbon dioxide transport; therefore, for the sake of simplicity, O_2 or CO_2 will not be included in subscripts and equations of general validity will be presented. Rather, subscripts m and b will be used to refer to free and chemically bound fractions of the dissolved gas, respectively. In the second part of the section, the analysis will be separately focused on O_2 or CO_2 and different equations will be introduced for the two gases.

With the above-described notation, the total concentration of a gas in blood is as follows:

$$c = c_m + c_b \qquad (6.13)$$

and the steady-state gas balance equation for gas transport may be written as:

$$\mathbf{v} \cdot \nabla (c_m + c_b) = \nabla \cdot (\mathscr{D}_m \nabla c_m + \mathscr{D}_b \nabla c_b) \qquad (6.14)$$

where the left-hand and right-hand side terms account for convective and diffusive transport of all species, respectively.

By assuming that the binding reactions are fast enough to ensure local equilibrium conditions, it is possible to write

$$\nabla c_b = H' \frac{\partial c_b}{\partial p} \nabla c_m \qquad (6.15)$$

Therefore, the continuity equation (Eq. 6.14) may be rewritten as

$$\mathbf{v} \cdot \left(1 + H' \frac{\partial c_b}{\partial p}\right) \nabla c_m = \nabla \cdot \left[\mathcal{D}_m + \mathcal{D}_b H' \frac{\partial c_b}{\partial p}\right] \nabla c_m \qquad (6.16)$$

The bracketed term in Eq. (6.16) is usually referred to as the *facilitated diffusion coefficient*, $\mathcal{D}_f$, that is,

$$\mathcal{D}_f = \mathcal{D}_m \left[1 + \frac{\mathcal{D}_b}{\mathcal{D}_m} H' \frac{\partial c_b}{\partial p}\right] \qquad (6.17)$$

Equation 6.17 shows that chemical binding results in an apparent enhancement of the molecular diffusivity of the dissolved gas; the *augmentation factor* (bracketed term in Eq. 6.17) quantifies the importance of facilitated diffusion and depends on the ratio of free to bound species diffusivity as well as on binding equilibrium conditions.

Finally, if a constant value can be assumed for $\partial c_b/\partial p$, Eq. 6.14 can be rearranged in terms of c_m only:

$$\mathbf{v} \cdot \nabla c_m = \mathcal{D}_{eff} \nabla^2 c_m \qquad (6.18)$$

where $\mathcal{D}_{eff}$ is the *effective diffusion coefficient,* which is defined as

$$\mathcal{D}_{eff} = \mathcal{D}_m \frac{1 + \dfrac{\mathcal{D}_b}{\mathcal{D}_m} H' \dfrac{\partial c_b}{\partial p}}{1 + H' \dfrac{\partial c_b}{\partial p}} = \frac{\mathcal{D}_f}{1 + H' \dfrac{\partial c_b}{\partial p}} \qquad (6.19)$$

6.6.2.1 Oxygen Transport

As reported in Sect. 6.6.1.1, dissolved oxygen is present in blood as a free molecular species and as oxygenated hemoglobin, which can be considered as oxygenated heme groups. Based on this consideration, Eq. 6.13 can be written for O_2 as

$$c_{O_2} = c_{O_2,m} + c_{HbO_2} \qquad (6.20)$$

where $c_{HbO_2} = 4 S_\% c_{Hb}$. According to Eqs. 6.13, 6.17, and 6.20, the O_2 facilitated diffusion coefficient can be written as

$$\mathcal{D}_{O_2,f} = \mathcal{D}_{O_2,m} \left[1 + \frac{\mathcal{D}_{HbO_2}}{\mathcal{D}_{O_2}} \frac{\partial c_{HbO_2}}{\partial c_{O_2,m}}\right] \qquad (6.21)$$

Since $\mathcal{D}_{HbO_2}/\mathcal{D}_{O_2} \ll 1$ due to the high hemoglobin molecular weight and concentration in red blood cells, the augmentation factor is significant only for very low oxygen partial pressures and can be neglected in physiological conditions (in which oxygen partial pressure is in the range of 40–95 mmHg). As a consequence, it can be assumed that $\mathscr{D}_{O_2,f} \simeq \mathscr{D}_{O_2,m}$ and the effective O_2 diffusivity is given by

$$\mathscr{D}_{O_2,eff} = \frac{\mathscr{D}_{O_2,m}}{1 + H'_{O_2} \dfrac{\partial c_{HbO_2}}{\partial p_{O_2}}} = \frac{\mathscr{D}_{O_2,m}}{1 + \mathscr{K} c_{Hb}} \tag{6.22}$$

6.6.2.2 Carbon Dioxide Transport

As already pointed out, several species originate from chemical binding of dissolved CO_2 in blood (see Sect. 6.6.1.2). According to the notation introduced, the total concentration of these species[13] will be denoted as $c_{CO_2,b}$ and CO_2 facilitated diffusion coefficient can be expressed as

$$\mathscr{D}_{CO_2,f} = \mathscr{D}_{CO_2,m} \left[1 + \frac{\mathscr{D}_{CO_2,b}}{\mathscr{D}_{CO_2,m}} \frac{\partial c_{CO_2,b}}{\partial c_{CO_2,m}} \right] \simeq \mathscr{D}_{CO_2,m} \left[1 + \frac{\mathscr{D}_{CO_2,b}}{\mathscr{D}_{CO_2}} \frac{H'_{CO_2}}{\mathscr{H}_{CO_2}} \right] \tag{6.23}$$

while CO_2 effective diffusivity is given by

$$\mathscr{D}_{CO_2,eff} = \frac{\mathscr{D}_{CO_2,f}}{1 + \dfrac{H'_{CO_2}}{\mathscr{H}_{CO_2}}} \tag{6.24}$$

6.7 Gas Exchange Between Capillary Blood and Alveolar Air

Before presenting and discussing mathematical models of blood oxygenators, a simple analysis of gas transfer in alveoli is reported in this section. Such analysis is aimed at understanding the results that blood oxygenators should obtain and why their performance is currently far from that of healthy lungs.

Let us consider gas exchange between blood in the alveolar capillary and air in the alveolar sac, as schematically represented in Fig. 6.7. Blood enters the venous end of the alveolar capillary with a gas (O_2 and CO_2, subscript omitted) pressure p_v; gas exchange occurs between blood and the alveolar sac, where the gas partial pressure is p_{alv}, as in the exhaled air. At steady state, the gas balance on an infinitesimal capillary segment along the axial direction z gives

---

[13]It is worth emphasizing once again that the bicarbonate ion HCO_3^- is by far the most abundant species originated by CO_2 reactions in blood; therefore, it is reasonable to assume $c_{CO_2,b} \simeq c_{HCO_3^-}$.

Fig. 6.7 Gas exchange between air in the alveolar sac and blood in the alveolar capillary

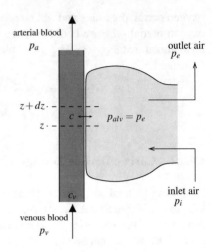

$$v\frac{\mathrm{d}c}{\mathrm{d}z} = -\frac{4}{d}N_{tm} \tag{6.25}$$

where v is the blood velocity, d is the capillary diameter, and N_{tm} is the transmembrane gas flux from blood to alveolar air. The flux N_{tm} can be written as

$$N_{tm} = K_c\left(c - c^*\right) \tag{6.26}$$

where K_c is the overall mass transfer coefficient through the respiratory membrane and c^* is the gas concentration in blood in equilibrium with the air in the alveolar sac.

By substituting Eq. 6.26 in Eq. 6.25, the following equation is obtained

$$\frac{\mathrm{d}c}{\mathrm{d}z} = -\frac{1}{Pe_{tm}}\left(c - c^*\right) \tag{6.27}$$

where the transmembrane Peclet number defined as

$$Pe_{tm} = \frac{vd}{4K_c} \tag{6.28}$$

was introduced. It is worth noting that Pe_{tm} may be equally defined as

$$Pe_{tm} = \frac{\dfrac{1}{aK_c}}{\dfrac{L}{v}} = \frac{Q_B}{AK_c} \tag{6.29}$$

where $a = 4/d$ is the specific exchange area per unit capillary volume, L is the capillary length, Q_B is the blood volumetric flow rate, and A is the total exchange area. Equation 6.29 highlights the physical meaning of Pe_{tm}, which is the ratio of the transmembrane diffusion time to the blood residence time in the capillary.

Equation 6.27 can be integrated with the boundary condition given by the known gas concentration at the venous end of the capillary, $c_{B,v}$; the solution obtained can be cast in the following dimensionless form

$$\tilde{c}(\tilde{z}) = 1 - \exp(-\tilde{z}) \tag{6.30}$$

where

$$\tilde{z} = \frac{z}{L\,Pe_{tm}} \quad ; \quad \tilde{c} = \frac{c - c_v}{c^* - c_v} \tag{6.31}$$

Figure 6.8 shows the plot of the dimensionless gas concentration in blood along the alveolar capillary (Eq. 6.30). From this dimensionless plot and accounting for the gas solubility in blood (see Sect. 6.6.1), it possible to determine the partial pressure profiles of oxygen and carbon dioxide along the capillary, which are shown in Fig. 6.9.

Equation 6.30 shows that the dimensionless gas concentration at the outlet ($z = L$, arterial end) of the alveolar capillary is $1 - \exp(-1/Pe_{tm})$; therefore, for low Pe_{tm} values, blood leaves the capillary with a gas concentration close to equilibrium with the alveolar air ($\tilde{c} \simeq 1$); in this case, it can be easily shown that the overall rate of gas exchange approaches the limiting value $vA\,(c^* - c_v)$.

Healthy human lungs operate with a low Q_B/A ratio (5 l/min of blood are spread over a surface of $100–150\,\mathrm{m^2}$); furthermore, the respiratory membrane offers a very

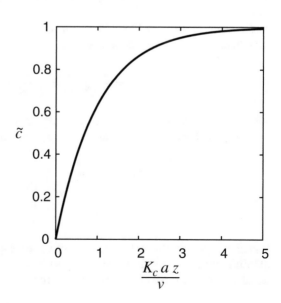

Fig. 6.8 Dimensionless gas concentration in blood along the alveolar capillary (Eq. 6.30)

Fig. 6.9 Oxygen and carbon dioxide pressure in blood along the alveolar capillary. Oxygen pressure is obtained with $p_v = 40$ mmHg, $p_{alv} = 105$ mmHg, and $c_{Hb} = 2.2$ mM; carbon dioxide concentration is obtained with $p_v = 45$ mmHg and $p_{alv} = 40$ mmHg

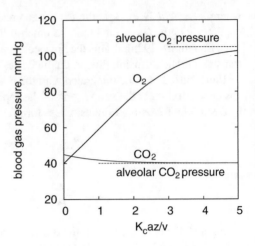

low resistance to gas transfer, so that the overall mass transport coefficient K_c depends only on the blood-side resistance and has a very high value. Due to both the low Q_B/A ratio and high K_c value, Pe_{tm} is low in the characteristic operating conditions of human lungs, so that gas pressure at the arterial end of the capillary and in the alveolar sac is virtually equal.

It is worth noting that the compositions of alveolar and inspired air are different, in fact, accounting for gas mass balance over the whole blood and alveolar compartments:

$$Q_B c_v + \frac{V_i}{\mathbb{R}T} p_i = Q_B c_a + \frac{V_i}{\mathbb{R}T} p_{alv} \tag{6.32}$$

where Q_B is the volume blood flow rate and V_i the volume flow rate of air entering the alveoli. The above equation may be rewritten as:

$$c_a - c_v = \frac{p_a - p_v}{\mathscr{H}} = \frac{V_i}{Q_B \mathbb{R}T}(p_i - p_{alv}) \tag{6.33}$$

where, according to Eqs. 6.8 and 6.12, $\mathscr{H}$ is set equal to the ratio $(p_a - p_v)/(c_a - c_v)$; by further assuming that $p_a \sim p_{alv}$, we get:

$$p_a = \frac{\dfrac{V_i \mathscr{H}}{Q_B \mathbb{R}T} p_i + p_v}{\dfrac{V_i \mathscr{H}}{Q_B \mathbb{R}T} + 1} \quad or \quad \frac{p_a - p_v}{p_i - p_v} = \frac{\dfrac{V_i \mathscr{H}}{Q_B \mathbb{R}T}}{\dfrac{V_i \mathscr{H}}{Q_B \mathbb{R}T} + 1} \tag{6.34}$$

Equation 6.34 underlines the influence of the ventilation–perfusion ratio V_i/Q_B on the difference between inspired and alveolar (or arterial) gas partial pressure. In healthy lungs at rest, $V_i/Q_B \sim 1$; therefore, as for blood oxygenation, using a mean $\mathcal{H}_{O_2}$ value of 27.8 mmHg mmole^{-1} l, we have $V_i \mathcal{H}_{O_2} / (Q_B \mathbb{R} T) \sim 1$ and $(p_{a,O_2} - p_{v,O_2}) / (p_{i,O_2} - p_{v,O_2}) \simeq 0.5$.

6.8 Gas Exchange in Membrane Oxygenator

Most of the oxygenators used in current clinical practice are based on hollow fiber membranes, with high exchange surface-to-volume ratio. Oxygen and carbon dioxide exchange occurs through the membrane, avoiding the direct contact between blood and sweep gas. From this point of view, such artificial devices try to mimic the natural operation of the lung; however, from a quantitative point of view, parameters such as total surface area, contact time, or membrane permeability are still not comparable (see Table 6.5). As a consequence, the performance of the currently available artificial devices is sufficient only for the support of the basal metabolic needs of patients.

The gas exchange rate may be written as:

$$F_{tm} = K_p A \left(p^G - p^* \right) \tag{6.35}$$

where K_p is the overall mass transport coefficient, A is the exchange surface area, p^* is the partial pressure of the gas in equilibrium with the blood, and p^G is the partial pressure in the sweep gas phase.

In order to determine the overall mass transport coefficient, the analysis reported in Chap. 2 must be extended to include also the mass transport resistance due to the membrane; we obtain:

$$\frac{1}{K_p} = \frac{1}{k_p^G} + \frac{\delta}{\mathscr{P}} + \frac{1}{k_p^B} \tag{6.36}$$

Table 6.5 Comparison of some operating variables in human lungs and membrane oxygenators

		Lung	Membrane oxygenators
Exchange surface area		$70\,m^2$	$1–3\,m^2$
Membrane material		Hydrophilic	Hydrophobic
	Gas permeability	High	Low
	Thickness	$1–2\,\mu m$	$50–100\,\mu m$
Gas		Air	Enriched air
Blood vessel	Diameter	$3–7\,\mu m$	$150–250\,\mu m$
	Length	$0.5–1\,cm$	$10–15\,cm$
	Contact time	$<1\,s$	$5–15\,s$

where $\mathscr{P}$ is the membrane permeability (see Chap. 3), δ is the membrane thickness, and k_p^G and k_p^B are the mass transport coefficients (referred to partial pressure as driving force) in the gas phase and blood phase, respectively.[14]

Usually, the resistance in the gas phase is negligible; as a consequence, the overall mass transfer coefficient is independent of gas-phase flow rate. In such a condition, reducing the gas-phase mass transfer resistance has a negligible effect on the trans-membrane gas flux; rather, in order to increase the gas transfer rate per unit surface of the device, all efforts should be aimed at improving the membrane permeance or reducing the resistance in the blood boundary layer.

Another way of increasing the transmembrane gas flow is by increasing the partial pressure driving force for gas exchange. This can be done quite easily for oxygen, by using enriched air or even pure oxygen (properly humidified) as gas phase: in this way, a driving force up to 500–600 mm Hg can be obtained. On the other hand, this is not as easily done for carbon dioxide, for which the maximum attainable driving force is 40–45 mm Hg, when a CO_2-free gas phase is fed to the device. Considering that oxygen and carbon dioxide exchange rates must be in the metabolic ratio (1:0.8), the mass transfer coefficient must be higher for carbon dioxide than for oxygen: more specifically, in order to exploit the maximum driving forces available, the CO_2 to O_2 mass transport coefficient ratio should be about 10. Such a result can be only be obtained by relying on the membrane resistance properties, since mass transport in blood phase is not significantly different for the two gases.

The following part of this section provides information on how to evaluate the mass transfer resistance of both the blood boundary layer and different types of membranes used in blood oxygenators.

6.8.1   Membranes for Gas Oxygenators

6.8.1.1   Dense Membranes

Polymeric membranes made of thin polymer sheets are often used for gas exchange in artificial lung, mainly for long-term extracorporeal life support.

---

[14]In detail, for mass conservation, the gas fluxes, through the gas phase, the membrane, and the blood phase must be equal, so that:

$$N_{tm} = k_p^G \left(p^G - p^{G,i} \right) = \frac{\mathscr{P}}{\delta} \left(p^{G,i} - p^{B,i} \right) = k_c^B \left(c^i - c \right) = \alpha_B k_c^B \left(p^{B,i} - p^* \right)$$

The mass transport coefficient in the blood, k_p^B, is then defined as

$$k_p^B = \alpha_B k_c^B$$

where α_B is the solubility of A in the blood (the solubility of A in the blood at the equilibrium with a gas phase with a partial pressure p is written as $c = \alpha_B p$).

Table 6.6 Permeability of various polymeric materials [59]

Polymer	$\mathscr{P}_{O_2} \dfrac{\text{mole mm}}{\text{m}^2 \text{min atm}}$	$\dfrac{\mathscr{P}_{O_2}}{\mathscr{P}_{CO_2}}$
Polyethylene	5×10^{-4}	5
Teflon	4×10^{-4}	3
Silicone rubber	5×10^{-2}	5

Gas permeation through a polymer layer is usually described by solution–diffusion model (see Chap. 3), which leads to the following expression for the membrane permeance:

$$\mathscr{P} = \alpha_m \mathscr{D}^m \tag{6.37}$$

where α_m is the gas partition coefficient in the membrane polymer.

The properties of some polymeric materials are reported in Table 6.6. Good results are obtained with 100–200 μm-thick silicone membranes, which allow to obtain mass transfer coefficients of the order of 0.05 mol min^{-1} m^{-2} atm^{-1} and exhibit excellent properties in terms of biocompatibility.

It is worth noting that while dense membranes have a lower permeability than the porous ones, so that a higher surface area and priming volume are required, they greatly reduce the risk of blood leakage or, conversely, gas entrainment into the blood; these properties as well as their longevity make them still the ideal candidate for long-term support.

6.8.1.2 Porous Membranes

Most of the membranes used in blood oxygenators are of the hollow fiber porous type, with fiber diameters of 200–400 μm and porous wall thickness of 20–50 μm; the fiber wall is highly porous (30–50 %) with pore size below 0.1 μm.

Hydrophobic polymers are used to prevent blood penetration into the membrane pores, which thus remain filled with gas. If θ is the contact angle between gas and blood on the polymer, the pressure required to fill the pores is $4\sigma \cos \theta / d_P$, where σ is the gas–blood surface tension and d_P is the pore diameter: For polymers that are not wetted by blood ($\theta \simeq \pi$) and small pore diameter, pore wetting is easily prevented, possibly by applying a small gas overpressure.

In empty pores, respiratory gases diffuse relatively fast and the permeability can be estimated by accounting for the membrane wall porosity (ε_w) and tortuosity factor (τ):

$$\mathscr{P}_{dry} = \frac{\varepsilon_w}{\tau} \frac{\mathscr{D}^P}{\mathbb{R}T} \tag{6.38}$$

Due to the small size of membrane pores, gas diffusion is likely to occur in the Knudsen regime (see Chap. 3) and diffusivity is given by $\mathcal{D}^P = 2d_p/3\sqrt{2\mathbb{R}T/\pi M}$, where M is the diffusing gas molecular mass; therefore, we get:

$$\mathcal{P} = \frac{2\varepsilon_w}{3\tau}d_p\sqrt{\frac{2}{\pi M\mathbb{R}T}} \tag{6.39}$$

According to Eq. 6.39, the ratio $\mathcal{P}_{dry}\delta$ (membrane permeance) can be as high as 50 mol min^{-1} m^{-2} atm^{-1} (2.5 × 10^{-2} ml cm^{-2}s^{-1} cmHg^{-1}); slightly lower values are obtained for carbon dioxide. With such high values, blood-side resistance controls the gas transfer process and $K_p \simeq k_p^B$.

On the other hand, the performance of porous membranes is limited by pore wetting issues: in fact, mainly when the membrane is used for prolonged support, adsorption of plasma components renders the membrane surface hydrophilic; in these conditions, plasma can enter and fill the pores. Therefore, the permeating gas has to diffuse in a plasma layer ($\mathcal{D}^P = \mathcal{D}$, where $\mathcal{D}$ is the diffusivity in plasma) and the permeability reduces to:

$$\mathcal{P} = \frac{\varepsilon_w \alpha_B}{\tau}\mathcal{D} \tag{6.40}$$

where α_B is the gas partition coefficient in blood. Pore wetting can cause a reduction of permeability up to 5 orders of magnitude compared to dry membranes, with a consequent degradation of the device performance.

6.8.1.3 Composite Membrane

In order to prevent wetting and infiltration into the microporous wall and improve the long-term performance of membrane oxygenators, composite hollow fibers are used, which incorporate a thin layer of non-porous polymer on the fiber surface. Composite fibers are produced either by coating a previously manufactured microporous membrane or by a one-step process in which both the porous and dense layers are formed at the same time.

The non-porous "skin" prevents plasma infiltration in the pores even during prolonged applications, but offers a further resistance to gas transport across the membrane. The non-porous skin may be considered as an additional membrane, and a term accounting for its mass transfer resistance should be included in Eq. 6.36. To that end, the permeance of the dense layer can be calculated as in Eq. 6.37. However, the non-porous skin is usually very thin, so that the gas exchange performance of composite and porous membranes is in general comparable.

6.8.2 Gas Transport in the Blood Film

The blood film near the membrane surface offers in general the controlling mass transfer resistance when a porous membrane is used. Therefore, improvement of blood fluid dynamics and mass transport coefficient is among the major aims of the current research on blood oxygenators.

As discussed in Chap. 2, the mass transport coefficient depends on the thickness of the boundary layer formed near the solid surface, which in turn depends on the blood velocity field. A decrease of the boundary layer thickness may be achieved by increasing the blood flow rate or reducing the cross-sectional area available to flow, with an appropriate device design.

Several studies have been carried out to evaluate the mass transport coefficient in blood oxygenators. A theoretical analysis based on the solution of the equation of motion in laminar flow is suitable only for simple geometries and flow conditions such as in intraluminal blood flow; differently, more complex flow patterns such as extraluminal flow require a different approach, which can involve experimental testing of commercial devices or 3D computational fluid dynamic simulations. Moreover, chemical binding of O_2 and CO_2 in blood strongly affects their transfer rates and should be accounted for.

The following approach is generally considered: firstly, the mass transport coefficient in a non-reacting system is determined (e.g., by experiments carried out by using water or simulated blood as liquid phase); then, the effect of binding reactions is accounted for.

As already reported in Chap. 2, the mass transport coefficient in a non-reacting system, k_p^{B0}, is usually described by correlations of the form:

$$Sh = \frac{k_p^{B0}\ell}{\alpha_B \mathscr{D}_B} = aRe^b Sc^{1/3} \tag{6.41}$$

Equation 6.41 shows that the mass transport coefficient depends on both the physical properties (Sc) and fluid dynamics of the system (Re). In Eq. 6.41, ℓ is a characteristic length for blood flow: for intraluminal flow, $\ell = d_f$, while for extraluminal flow, ℓ is the equivalent diameter given by $\varepsilon_f d_{fe}/(1 - \varepsilon_f)$; d_f and d_{fe} are the inner and outer diameters of the hollow fibers, respectively, and ε_f is the void fraction of the fiber bundle. Some correlations for the blood-side mass transport coefficient in different flow configurations are summarized in Table 6.7.

Two different approaches have been suggested in the literature to account for the enhancement of the gas transfer rate due to the binding reactions.[15]

A first approach [65, 66] uses the same relations derived for non-reacting system, but accounts for facilitated and effective diffusion as defined in Sect. 6.6.2. Specifically, it was proven that the effective diffusivity should be used in calculating the Schmidt number for blood, whereas the Sherwood number is based on the facilitated

---

[15]Many of the literature works focus on the oxygen transfer rate; few works deal also with carbon dioxide transfer. However, the same approach can be followed for both gases.

Table 6.7 Correlations for the blood-side mass transfer coefficient

Geometry	Correlations	Ref	Note
Flow outside and across bundles of hollow fibers	$Sh = 0.8Re^{0.47}Sc^{0.33}$	[60]	
	$Sh = 0.8Re^{0.59}Sc^{0.33}$	[60, 61]	a
	$Sh = Re^{2/3}Sc^{1/3}\exp\left(3.26\varepsilon_f - 4.27\right)$	[62]	b
	$Sh = 0.46Re^{0.76}Sc^{0.33}$	[63]	
Cross-flow outside to hollow fiber mat	$Sh = 0.39Re^{0.76}Sc^{1/3}$	[64]	c
	$Sh = 0.52Re^{0.623}Sc^{1/3}$	[65]	d
Flow in thin channels (slit)	$Sh = 0.5Gz$	[60]	e

[a]The correlation has been verified also for non-Newtonian fluids, using properly defined Reynolds and Schmidth numbers.
[b]Cross-wound hollow fibers; ε_f is the void fraction $(0.39 \div 0.65)$, $\ell = \varepsilon_f / \left(1 - \varepsilon_f\right) d_{fe}$, where d_{fe} is the outside diameter of the fiber
[c]From computational and experimental studies, both with steady and pulsatile flow
[d]From numerical calculation
[e]$Sh = k_B\left(4B\right)/\mathscr{D}$, Graetz number $Gz = (4B)^2 u/\mathscr{D}L$, where B is the half-thickness and L is the length of the channel

diffusion. Following this approach, the ratio of the oxygen transport coefficient in blood to its transport coefficient in a non-reacting system is given by (the augmentation factor in the facilitated diffusivity is negligible for oxygen):

$$\frac{k_B}{k_B^0} = \left[1 + \frac{4c_{\mathrm{Hb}}\left(\mathrm{d}S_{\%}/\mathrm{d}p_{O_2}\right)}{\alpha}\right]^{1/3} \tag{6.42}$$

In a second approach, the effect of binding reaction on the mass transport coefficient is quantified by multiplying the mass transport coefficient determined for a non-reacting system, k_B^0, by an enhancement factor E:

$$k_B = k_B^0 E$$

As for oxygen, the enhancement factor can be evaluated from experimental data [67] or from a theoretical analysis based on the film model [68].

Relations to evaluate the enhancement factor are reported in Table 6.8.

6.9   Membrane Oxygenator Modeling

A reliable mathematical model can be considered as the basis for a rational design of a blood oxygenator, as for any other device; furthermore, mathematical models can be a valuable tool to get insight into the working principles of oxygenators,

Table 6.8 Enhancement factor for oxygen mass transport coefficient

Correlations	Ref	Note
$E = \left(\dfrac{95000}{p_{O_2}^G}\right)^{1/3} \times \left[1 + 11.8\left[(1 - S_\%)\dfrac{\text{Htc}}{100}\right]^{0.8} - 8.9\left[(1 - S_\%)\dfrac{\text{Htc}}{100}\right]\right]$	[62]	a
$E = 1 + \dfrac{n\mathscr{D}_{RBC}}{\mathscr{D}_{O_2}} \dfrac{1}{1 + K_E\left(c_{O_2}^{B,i}\right)^n} \dfrac{\left(c_{O_2}^{B,i}\right)^n - \left(c_{O_2}\right)^n}{c_{O_2}^{B,i} - c_{O_2}} K_E c_{Hb}$	[68]	b

[a] Determined from the comparison of mass transport coefficient in normal blood and in blood where hemoglobin was inactivated by carbon monoxide; Htc is the hematocrit (%)
[b] Determined from film model, accounting for Hill equation for oxygen–hemoglobin reaction. $\mathscr{D}_{RBC}$ red blood cell diffusivity, $K_E = \left(H'_{O_2}/p_{50}\right)^n$

by helping in the definition of the critical design parameters and of their effect on the performance, also without having to carry out extensive, costly, and complex experimental campaigns.

Though more sophisticated and detailed models can be developed at the price of a greater mathematical and numerical complexity, in this section we present simplified models based on reasonable hypotheses. By providing a fair trade-off between simplicity and reliability of analysis, such models allow to focus only on the fundamental aspects of the oxygenation process, rather than on involved mathematical solutions.

The model presented here is based on the following assumptions: (a) steady-state conditions; (b) the oxygen–hemoglobin reaction is always at equilibrium; (c) the linearized form of the oxygen saturation curve holds; (d) gas transport in both bulk phases is purely convective; (e) the active volume (i.e., the space where blood and gas are indirectly contacted through the membrane and gas exchange occurs) of the device can be considered as a porous pseudo-continuous medium, through which blood and gas flow; and (f) the gas transfer rate between gas and blood is given as $k_p^B a \left(p^G - p^*\right)$, where a is the membrane surface area per unit total volume.

Under the above-listed hypotheses, the oxygen balance in blood may be written as:

$$\mathbf{v}_B \cdot \nabla \left(c_{O_2,m} + c_{HbO_2}\right) = K_{p,O_2} a \left(p_{O_2}^G - p_{O_2}^*\right) \tag{6.43}$$

where $\mathbf{v}_B$ is the blood superficial velocity and a is the membrane surface area per unit total volume. The above equation may be rearranged in the form

$$\mathbf{v}_B \cdot \frac{1 + \mathscr{H} c_{Hb}}{H'_{O_2}} \nabla p_{O_2}^* = \mathbf{v}_B \cdot \frac{\nabla p_{O_2}}{\mathscr{H}_{O_2}} = K_{p,O_2} a \left(p_{O_2}^G - p_{O_2}^*\right) \tag{6.44}$$

As for the gas phase, the oxygen balance is written as

$$\frac{\mathbf{v}_G}{\mathbb{R}T} \cdot \nabla p_{O_2}^G = -K_{p,O_2} a \left(p_{O_2}^G - p_{O_2}^*\right) \tag{6.45}$$

where $\mathbf{v}_G$ is the gas superficial velocity.

In the following sections, this general model is solved for two common flow patterns used in blood oxygenators.

6.9.1   Countercurrent Blood Oxygenator

In countercurrent flow, blood and gas flow in opposite sense along the same direction and exchange oxygen and carbon dioxide across the membrane (see Fig. 6.10). This flow pattern is found in hollow fiber or flat-sheet oxygenators, and the simple analysis presented here applies equally to both types of device.

Let x be the direction of flow, and then both gas and blood have velocities directed along x; furthermore, both blood and gas compositions vary with x, due to mass transfer across the membrane. The oxygen balance equations on an infinitesimal segment of the device along the x direction can be written as

$$\frac{v_B}{\mathcal{H}_{O_2}} \frac{d p_{O_2}^*}{dx} = K_{p,O_2} a \left(p_{O_2}^G - p_{O_2}^* \right) \tag{6.46}$$

$$\frac{v_G}{\mathbb{R}T} \frac{d p_{O_2}^G}{dx} = -K_{p,O_2} a \left(p_{O_2}^G - p_{O_2}^* \right) \tag{6.47}$$

The above set of differential equations can be solved with the boundary conditions:

$$\begin{cases} x = 0 & p_{O_2}^* = p_v \\ x = L & p_{O_2}^G = p_{in,O_2}^G \end{cases}$$

where p_v is the oxygen pressure in the inlet venous blood and p_{in,O_2}^G is the oxygen partial pressure in the inlet gas. By some simple mathematical rearrangements and integration via variable separation, we get:

$$\ln \frac{p_{out,O_2}^G - p_v}{p_{in,O_2}^G - p_a} = K_{p,O_2} a L \left(\frac{\mathbb{R}T}{v_G} - \frac{\mathcal{H}_{O_2}}{v_B} \right) \tag{6.48}$$

or

$$\ln \frac{p_{out,O_2}^G - p_v}{p_{in,O_2}^G - p_a} = K_p A \left(\frac{\mathbb{R}T}{V_G} - \frac{\mathcal{H}_{O_2}}{Q_B} \right) \tag{6.49}$$

Fig. 6.10 Schematic geometry of countercurrent flow oxygenators

where p_a is the oxygen pressure in the arterial blood, p_{in}^G and p_{out}^G are the oxygen pressure in the inlet and outlet gas, respectively, A is the total membrane area, and V_G and Q_B are the gas and blood volumetric flow rates, respectively.

An oxygen balance over the whole device allows to obtain the following expression for the overall oxygen transfer rate in the device

$$F_{tm,O_2} = \frac{V_G}{\mathbb{R}T} \left(p_{in,O_2}^G - p_{out,O_2}^G \right) = \frac{Q_B}{\mathscr{H}_{O_2}} (p_a - p_v) \tag{6.50}$$

By substituting Eq. 6.50 in Eq. 6.49, the following expression is obtained

$$F_{tm,O_2} = K_{p,O_2} A \frac{\left(p_{out}^G - p_v \right) - \left(p_{in}^G - p_a \right)}{\ln \dfrac{p_{out}^G - p_v}{p_{in}^G - p_a}} \tag{6.51}$$

$$F_{tm,O_2} = K_{p,O_2} A \, \Delta p_{LM,O_2} \tag{6.52}$$

In Eq. 6.52, the *log mean partial pressure difference*, $\Delta p_{LM,O_2}$, was introduced. It is worth noting that Eq. 6.51 is valid also when it is possible to assume that the partial pressure of oxygen in the gas stream does not vary significantly across the device (i.e., $p_{in,O_2}^G \simeq p_{out,O_2}^G = p_{O_2}^G$); such condition holds if $V_G/\mathbb{R}T \gg Q_B/\mathscr{H}_{O_2}$. Furthermore, Eq. 6.51 applies also to cocurrent flow, i.e., when gas and blood flow in the same sense inside the oxygenator; however, such flow pattern is not used in clinical devices because of its lower efficiency compared to countercurrent flow.

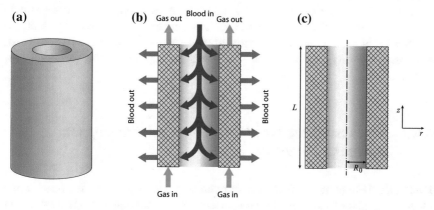

Fig. 6.11 Schematic geometry of radial flow oxygenators; **a** perspective view, **b** cross-sectional view with flow pattern, **c** cross-sectional view with sizes

6.9.2 Hollow Fiber Oxygenator with Radial Blood Flow

Another flow pattern used in membrane oxygenators, blood flows radially (direction r) across bundles of woven hollow fibers, while gas flows in the fiber lumen in the axial (z) direction as schematically described in Fig. 6.11 (see also Fig. 6.4). In such configuration, the velocity of blood is not constant, but varies along the flow direction as follows:

$$v_B(r) = \frac{Q_B}{2\pi L r} \tag{6.53}$$

The local oxygen balances in the blood and gas sides are as follows:

$$\frac{Q_B}{2\pi L \mathscr{H}_{O_2}} \frac{1}{r} \frac{\partial p_{O_2}^*}{\partial r} = K_{p,O_2} a \left(p_{O_2}^G - p_{O_2}^* \right) \tag{6.54}$$

$$\frac{v_G}{\mathbb{R}T} \frac{\partial p_{O_2}^G}{\partial z} = -K_{p,O_2} a \left(p_{O_2}^G - p_{O_2}^* \right) \tag{6.55}$$

with the boundary conditions:

$$\begin{cases} r = R_0 & p = p_v \\ z = 0 & p^G = p_{in}^G \end{cases}$$

where R_0 is the inner radius of the active space of the device, i.e., where blood is fed in the external space of the fiber bundle.

It is worth noting that, in Eqs. 6.54 and 6.55, $p_{O_2}^G$ and $p_{O_2}^*$ are functions of both r and z. The above set of differential equations can be solved numerically. A closed solution is obtained if the oxygen partial pressure in the gas phase can be considered as constant; in this case, it can be shown that Eq. 6.52 holds also for radial flow devices.

6.9.3 Response of the Membrane Oxygenator to Different Operating Conditions

Equation 6.51 may be rewritten in a more useful form, to analyse the response of a blood oxygenator to changes in the operating conditions, such as gas flow rate or inlet gas composition. To this aim, it is useful to define two-dimensionless groups

$$R = \frac{K_p A}{Q_B/\mathscr{H}} \qquad Y = \frac{V_G/(\mathbb{R}T)}{Q_B/\mathscr{H}} \tag{6.56}$$

The first one, R, accounts for the ratio of the mass transfer rate to blood convection rate: a low value of R means that the gas transfer is slow compared to the rate at which blood passes through the device; therefore, when R is low, gas transfer tends to be ineffective. The second parameter, Y, accounts for the gas to blood flow rate ratio. Equations 6.50 and 6.51 may be rearranged to get:

$$\frac{F_{tm}}{\frac{Q_B}{\mathcal{H}}\left(p_{in}^G - p_v\right)} = \frac{p_a - p_v}{p_{in}^G - p_v} = \frac{Y\left[1 - \exp\left[-R\left(1 - Y\right)/Y\right]\right]}{1 - Y\exp\left[-R\left(1 - Y\right)/Y\right]} \quad (6.57)$$

The ratio $(p_a - p_v)/(p_{in}^G - p_v)$ represents an effectiveness factor for the gas transfer in the device, i.e., the ratio between the gas rate effectively transferred to the blood (proportional to the difference between the gas pressure in the arterial and venous blood) and the maximum gas transfer rate using a sweep gas at p_{in}^G (proportional to the difference between the pressure in the sweep gas and the gas pressure in the venous blood); $(p_a - p_v)/(p_{in}^G - p_v) = 1$ corresponds to the maximum transfer rate achievable in the device, with an outlet blood in equilibrium with the inlet gas; on the other hand, when $(p_a - p_v)/(p_{in}^G - p_v) = 0$, gas transfer does not occur.

Figure 6.12 reports the plot of $(p_a - p_v)/(p_{in}^G - p_v)$ as a function of R and Y. From the plot, it is evident that for low Y values—i.e., for low values of the ratio between gas and blood flow rates—the gas transfer rate to blood increases approximately linearly with Y, but becomes relatively constant once Y is sufficiently high; for $Y \to \infty$, the ratio $(p_a - p_v)/(p_{in}^G - p_v)$ approaches the limiting value of $1 - e^{-R}$. Therefore, if the goal is to ensure the maximum possible rate of gas

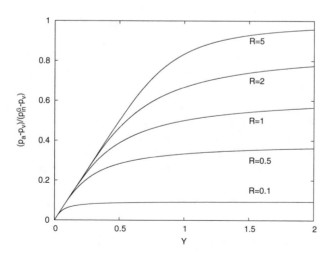

Fig. 6.12 Effectiveness factor of the gas transfer in the oxygenator as a function of the two operating parameters R and Y

exchange, we need to operate above the minimum gas flow rate that corresponds to a gas transfer rate independent of the sweep gas flow; on the other hand, operating at lower Y values, it is possible to control the gas exchange rate by manipulating the sweep gas flow rate.

6.10   Current Research and Perspectives

The artificial lungs used today in the clinical practice are extracorporeal devices, mainly used in open-heart surgery to replace the heart and lung functions; in this case, the device is required to ensure gas exchange for some hours. The patient is anticoagulated with heparin to prevent thrombosis within the extracorporeal circuit and potential formation of thromboemboli; the whole bypass circuit is often coated with heparin to prevent clotting and reduce the amount of systemic anticoagulant required.

Current research is devoted to finding solutions also for different applications, such as providing support in acute respiratory failures (such as ALI or ARDS), treating chronic respiratory diseases (such as COPD), and bridging to organ transplantation or allowing the natural lungs to recover; solutions for efficient and safe longer-term support (from several days to months) are then required. The traditional approach based on mechanical ventilation has some major drawbacks: the positive airway pressures and volume excursions associated with mechanical ventilation can cause further damage to the lung tissue, including barotrauma (high airway pressures), volutrauma (lung distension), and parenchymal damage due to the toxic levels of oxygen required for effective mechanical ventilation. Despite advances in support-ive care, the mortality rate in patients with the acute respiratory distress syndrome (ARDS) is widely considered to have remained high and generally in excess of 50 %. Therefore, the idea is to provide breathing support independent of the lungs, using respiratory assist devices as alternatives or adjuvants to mechanical ventilators for patients with failing lungs.

Lung assists are usually classified in (a) extracorporeal devices, not too different from those currently used in heart surgery, (b) paracorporeal or wearable devices that can be attached directly to the patients, and (c) intracorporeal or implantable devices that can be implemented with the intravenous or intratoracic configuration.

This section presents an overview of the recent developments in the field of arti-ficial lungs.

6.10.1   Extracorporeal Lung Assist (ECLA)

Extracorporeal lung assist is aimed at allowing the lungs to rest and recovery or bridging the patient to lung transplantation. In this framework, it is interesting to define two specific therapeutic approaches [69]

- total ECLA when both blood oxygenation and CO_2 removal are required;
- partial ECLA which aims principally to CO_2 removal, possibly with mild oxygenation to support natural lungs or low-pressure ventilation.

ECMO is suitable for total ECLA, providing both extracorporeal oxygenation and carbon dioxide removal. It is important to recognize that oxygenation is controlled by hemoglobin saturation[16] and effective oxygenation can be obtained only with high blood flow rates (at least 50–60 ml/min per kg of body weight). In ECMO, a high blood flow rate (70–80 % of the cardiac output) is diverted to a pump-driven external circuit including a membrane oxygenator and a heat exchanger to control the body temperature. The system is then similar to a cardiopulmonary bypass circuit, which demands a continuous bedside management of trained staff. A large membrane surface area (a square meter of hollow fiber membrane) is currently required to provide adequate gas exchange; furthermore, respiratory support may be required also for relatively long periods of time; therefore, in extracorporeal circuits, blood/biomaterial contact is extensive and the inflammatory or thrombogenic complications are exacerbated; finally, the membrane device is required to maintain its performance for a long period and resist to plasma wetting, which causes a decrease of permeability.

Research on extracorporeal devices for prolonged use is now focused on the enhancement of gas exchange and, therefore, on the reduction of the surface area of the membrane; improvement of the biocompatibility of the materials used is also a major target. In parallel, efforts to simplify the circuit device system and reduce the need for intensive monitoring are carried out. In order to increase the gas exchange effectiveness, improvements of fluid dynamics have been considered, which involve optimization of the design of the gas exchanger or active mixing of blood.

In contrast to the high blood flow required for blood oxygenation, the removal of metabolic CO_2 can be obtained also by treating a low blood flow rate (less than 25 % of the cardiac output) with a less invasive procedure. This is the basis for implementing a low flow technique for extracorporeal CO_2 removal (ECCO2R), while oxygenation remains a function of natural breathing. Blood flow in ECCO2R circuit can be pump-driven veno-venous or pump-less arteriovenous.

Pump-driven veno-venous ECCO2R devices use a blood flow rate depending on the clinical demands (see Table 6.9 for an example) up to a high-flow ECMO, if a suitable catheter is used.

A very small membrane area ($0.3\,m^2$) is used in the decap® system (Hemodec, Salerno, Italy): in this device, a recirculation loop of ultrafiltrate produced by a hemofilter—in series with the membrane oxygenator—increases the flow rate in the membrane module and enhances carbon dioxide removal (see Fig. 6.13). In a similar system (Decapsmart, Medica, Medolla, Italy), the CO_2 removal is combined to renal replacement therapy (RRT) in the patient with multiple organ failure needing both respiratory and renal support (see Fig. 6.14) [70].

ALung Technologies, Inc. commercializes the Hemolung® Respiratory Assist System, a dialysis-like alternative or supplement to mechanical ventilation originally

---

[16]Venous blood can only be loaded with 40–60 Nml/L of oxygen before hemoglobin saturation.

Table 6.9 Features of the iLA active® system by Novalang

	Low flow	Mid flow		High flow
CO_2 removal	partial	yes		yes
Oxygenation	no	Partial		yes
Access	VV	VV		VV
Invasiveness	Low	Moderate		High
Membrane device	MiniLung® petite	MiniLung®	iLA®	XLung®
Membrane area, m^2	0.32	0.65	1.3	1.9
Blood flow rate, ml/min	<0.8	<2.4	0.5–4.5	1–3
Priming volume, ml	55	95	175	275

Fig. 6.13 Scheme of the decap® system

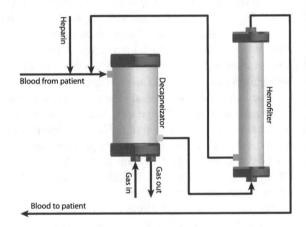

developed at the University of Pittsburgh. The primary component of the Hemolung® system is a cartridge which houses a cylindrical bundle of hollow fiber membranes; the fibers are positioned around a spinning core, which simultaneously drives blood flow (centrifugally) through the cartridge and CO_2 transfer from blood to the oxygen sweep gas flowing under negative pressure through the fiber lumen (see Fig. 6.15). The pump-driven flow past the membranes markedly increases the gas exchange efficiency, allowing for a significant CO_2 removal (30 to 40 %) at a relatively low blood flow in the range of 300 to 500 ml/min. Such a blood flow rate, similar to that used in hemodialysis, can be obtained with a single double-lumen venous catheter. A priming volume of 300 ml and minimal heparinization are required [72, 73].

Arteriovenous carbon dioxide removal ($AVCO_2R$) is carried out with a membrane gas exchange device connected directly from arterial to venous circulation, without a blood pump; a fraction of the cardiac output (10–30 %), dictated by the arterial to venous pressure difference and hydraulic resistance of the device, is diverted to the membrane unit, which operates with a high ventilation ratio. In this way, the device provides sufficient gas exchange to achieve nearly total removal of CO_2 and provides for about 10 % of the O_2 requirement. Oxygenation is then maintained by simple dif-

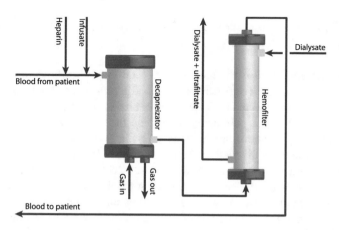

Fig. 6.14 Scheme of the decapsmart® system

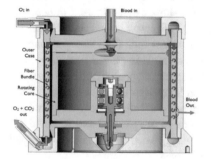 

Fig. 6.15 Hemolung® cartridge. *Left* cross-sectional drawing (reproduced from [71], under the CC BY 2.0 license—http://creativecommons.org/licenses/by/2.0/). *Right* picture of the filled cartridge (© 2013 ALung Technologies, Inc.)

fusion across the patient's alveoli and/or reduced mechanical ventilation. Even if a standard membrane unit can be used, it is of paramount importance to minimize the pressure drop in the device: computational fluid dynamics offers a major help to proper design the membrane module with pressure drops of few mmHg. An example of these devices is the interventional lung assist (iLA) of Novalung (GmgH, Hechineng, Germany), which uses a membrane "ventilator" with arteriovenous femoral connection.

6.10.2 Paracorporeal or Intrathoracic Devices

Extensive research is devoted to develop totally artificial lungs able to fully support basal O_2 and CO_2 transfer requirements for long-term support in chronic respiratory failure. The ultimate goal is to obtain an implantable device that can be placed inside body cavities; nevertheless, the present-day prototypes are implemented and tested in a paracorporeal configuration, with the device placed externally to the patient's chest.

The fundamental idea is to attach the artificial lung directly to the pulmonary circulation, by utilizing the right heart as the blood pump. In-series (with both outflow and inflow cannula connected to the pulmonary artery) and in-parallel (with outflow cannula connected to the pulmonary artery and the inflow cannula connected to the left atrium) applications are possible (see Fig. 6.16). In the former case, the whole cardiac output is diverted to the gas exchange device and the natural lungs act as embolic filter; however, the load on heart increases because of the hydraulic resistance offered by the device. In the case of in-parallel application, only a fraction of the cardiac output is diverted to the natural lung and receives respiratory support; the blood flow rate fed to the device depends on the relative hydraulic resistance of the artificial lung compared to the natural lung. This configuration requires a lower load for the heart. Recently, a new cannula design (Wang-Zwische double-lumen cannula, or W-Z DLC) based on the double-lumen cannula used in neonatal and pediatric veno-venous ECMO was also used. This new cannula replaces the in-series and in-parallel anastomoses with a single access including two pathways: a drainage pathway and an infusion pathway. The cannula is placed percutaneously through the internal jugular vein; the drainage lumen is open to both the superior and inferior vena cava, while the infusion lumen is open to the right atrium. Blood from systemic circulation flows through the superior and inferior vena cava into the drainage lumen to the artificial lung device. Blood is oxygenated and returned via the infusion lumen into the right atrium. This oxygenated blood is then pumped

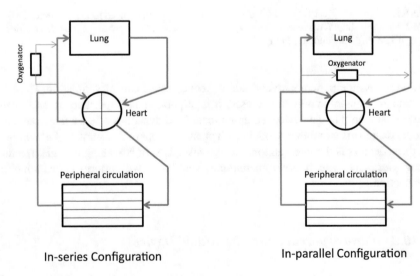

In-series Configuration   In-parallel Configuration

Fig. 6.16 In-series and in-parallel application of paracorporeal oxygenators

through the native circulation and pulmonary bed, thus receiving the full metabolic and filtering capacities of the native lungs [74].

It is important to consider that virtually all standard devices for heart–lung machines with hollow fiber membranes could be used for paracorporeal applications, but important technical improvements are required to obtain efficient and safe devices. Firstly, while pressure drop across the fiber bed is only of secondary importance for heart–lung machines equipped with external pumps, this parameter should be kept as low as possible when the right ventricle is used to drive blood circulation into the device. As an example, BioLung® from MC3 Inc. (Ann Arbor, MI, USA) utilizes radial flow through a concentrically wound hollow fiber fabric; computational fluid dynamics has been utilized to optimize the device, and the pressure drop has been reduced to 5–10 mmHg with a blood flow rate of 4–6 l/min and a membrane area of 1.5–2 m^2 [75].

In order to reduce the required surface area of membrane and the size of the devices, active mixing with a rapidly rotating disk is introduced in chronic artificial lung (CAL), developed by the University of Maryland [76]; mixing improves gas exchange and the centrifugal motion enables to pump the blood and to reduce the impact on the right heart in the in-series configuration.

Haemair Ltd. (Swansea, UK) patented a prototype of a portable lung, to be used by conscious, mobile patients [77]. Such system is based on a compact gas exchange device with high membrane surface area (5–20 m^2), still lower than the surface of natural lung but much larger than conventional oxygenators. The large surface area is required since the device is designed to use air instead of oxygen (thus avoiding the need of an oxygen supply like a bulky oxygen cylinder); furthermore, the idea to use the device for long-term conscious, mobile patient increases the oxygen demand above the basal requirement. A control system for sensing the patient demand for oxygen and adjust the blood and/or air flow is included: more specifically, the sensor detects the pulse rate of the patient, which is related to the oxygen demand.

The medium-term development of the devices presented here is aimed at obtaining small, implantable mass transfer devices, with blood flow driven by the natural circulation and supported by a small air pump placed outside the body. In the long-term, the gas exchange device should be included within a complete prosthetic lung that will employ no electrical or mechanical parts: the natural lungs should be removed and replaced with an elastic air sac placed in the pleural cavity; the natural breathing action should expand and contract the air sac, drawing air through the mass exchange apparatus.

6.10.3 Intravenous Devices

Intravenous gas exchange represents an attractive modality to support the respiratory function in ARDS. The fundamental idea is to transfer oxygen to and to remove carbon dioxide from venous blood with a bundle of hollow fiber inserted through a peripheral vein and placed in the vena cava, without requiring extracorporeal circu-

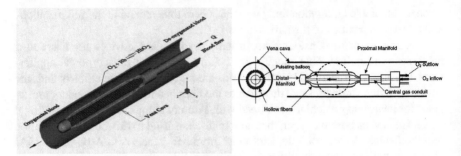

Fig. 6.17 Scheme of the IMO device with the *hollow* fiber bundle and the central *balloon* used to improve blood-side gas transport (reproduced from [80], with permission)

lation. Obviously, the fiber bundle must be compact to be safely placed in the vena cava, which has a diameter of 1.5–3 cm. This clearly introduces very strong constraint on the maximum allowable membrane area, so that the device can only support the gas exchange function, without replacing it. Generally, a respiratory support at 50 % of the basal requirement is considered as an appropriate target for these devices.

The first prototype of intravascular artificial lungs dates back to the 1980s with the IVOX [78] at CardiPulmonics, Inc. (Salt Lake City, UT, USA). Ivox consisted of a bundle of crimped hollow fiber membranes connected to a dual-lumen gas conduit which led outside the body to a console for providing gas flow through the fibers. The crimped fibers helped to minimize fiber clumping and to disturb blood flow improving the gas exchange permeance. The device was tested in clinical trials with a membrane area of 0.2–0.5 m^2 and providing for about 30 % of basal gas exchange requirements; the trial demonstrated the possibility to insert a fiber bundle in the vena cava for a prolonged periods without hemodynamic complications or thrombus formation; however, the extent of respiratory support was considered insufficient.

The McGowan Institute for Regenerative Medicine at the University of Pittsburgh has been active in the development of a respiratory support catheter usually referred to as intravenous membrane oxygenator (IMO) or Hattler catheter [79]: like IVOX, IMO consists of a bundle of hollow fibers, but a central polyurethane balloon which rhythmically (300 beats/min) inflates and deflates and provides blood mixing, thus enhancing the gas transfer coefficient (see Fig. 6.17). Flow velocity profiles visualized in the laboratory have shown that balloon pulsation also disrupts the layer of fluid near to the vessel wall, where the shear stress is higher, thus reducing the hemolytic damage. Tests for O$_2$ and CO$_2$ exchange in cow both ex-vivo and in vivo show that the balloon pulsation results in a significant increase in gas exchange and allows to meet the design target.

6.10.4   Microfluidic Devices[17]

Upon recognizing that the main limitations of ECMOs (even in the case of recent commercial equipment, such as the Maquet Quadrox) are ultimately to be ascribed to limited surface-to-volume ratio of the membrane module, research has been oriented toward scaling down the membrane exchange module to the microfluidic domain, i.e., shrinking down the characteristic dimension of blood- and gas-hosting channels from the current values of order 500 μm to channels of characteristic dimension of order 10–20 μm. Likewise, microfabrication techniques allow to reduce the thickness of the membrane separating blood and gas from the current value of order 50–100 μm to membranes as thin as 1–10 μm.

Such reduction of characteristic lengths makes it sensible to predict that ECMO devices exploiting this technology should eventually overcome the limitations of current hollow fiber technology by increasing the surface-to-area ratio up to two orders of magnitude and decreasing membrane resistance by an order of magnitude or even more, while maintaining the same or even decreasing the priming volume. Furthermore, because of the increased efficiency and lower flow rates involved by downsizing the characteristic cross-sectional length scales, atmospheric gas feeding and lower pressure drop can be afforded, thus allowing blood flow to be driven by heart pumping and the membrane exchanger to be fed by room air.

A quantitative assessment of the potential benefit of exploiting microfluidic technology in ECMO has been recently put forward by Potkay, who set up a simple transport model to estimate oxygen exchange rate per unit surface in a microfluidic device where the blood and the gas mixture are arranged in a multichannel cross-flow configuration [50, 81]. A conceptual scheme of the portable device together with the flow configuration for the membrane module is depicted in Fig. 6.18. This model provides a useful prediction of the average rated flow, Q_B/A, defined as the maximum blood flow rate for a given surface area of the membrane exchanger that allows inlet blood at oxygen saturation of 70 % to be collected at saturation of 95 % at the module outlet. Note that these values are fixed by physiological constraints. Quantitatively, Potkay proposes

$$\frac{A}{Q_B} = \alpha_B \frac{1}{K_{p,O_2}} \log \frac{p^*_{in,O_2} - p^G_{O_2}}{p^*_{out,O_2} - p^G_{O_2}} \tag{6.58}$$

The model expressed by Eq. 6.58 allows to estimate the potential limits of ECMO exploiting the microfabrication/microfluidic approach. For instance, in the case where the characteristic length of the channel cross-section is set to 10 μm and assuming that the membrane resistance to oxygen transport can be neglected, the predicted rated flow per square meter membrane area approaches 27 l min^{-1} [81], a value that overcomes by a factor five the best currently available commercial ECMOs.

---

[17]This section was authored by Stefano Cerbelli, Department of Chemical Engineering Materials and Environment, University "La Sapienza" of Rome.

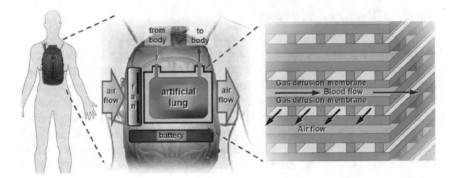

Fig. 6.18 Conceptual scheme of a portable lung support device, together with the flow configuration for the membrane module as proposed by [50] (reproduced with permission)

However, there are still many technological and even theoretical issues that prevent the achievement of this theoretical limit. Among the first category, one can single out the necessity for an accurate design of the microfluidic channel network, which should minimize pressure drop, thus allowing natural blood pumping through the device.

In this regard, natural systems still provide the most useful source for inspiration, such as in Murray's seminal work carried out in the mid-twenties of the last century, which established the minimum work principle in physiological flows [82, 83]. Murray's principle fixes a rule for the diameter ratio and the angle that should be satisfied when a larger channel branches into two smaller channels. In this respect, the possibility of integrated computer-assisted design of the channel network joint with soft lithography techniques could provide the practical chance to test these ideas at length scales that closely mimic those of natural vascular systems. Note, however, that an ideal microfluidic ECMO should contain several thousands of branched channel networks running in parallel in order to achieve suitable exchange surface area, an occurrence that makes the practical implementation of device microfabrication considerably more troublesome than the simple configuration devised by Potkay. Further practical challenges come from biocompatibility issues associated with the materials that come into contact with blood. Materials that are currently used to prevent thrombogenesys and platelet activation (e.g., heparin and PMEA) allow device lifetimes that are measured out in weeks.

Among the practical challenges so far described, which prevent ECMO from being a completely clinically successful technique, one is likely to be met in the near future, i.e., that of membrane thickness. New fabrication techniques such as initiated chemical vapor deposition (iCVD) are indeed continuously being proposed and tested, which are pushing membrane thickness below the single micrometer scale. To this end, it is worth remarking how iCVD has been shown to be compatible with branching channel network geometries [84]. Figure 6.19 shows an example of application of this technique to a branched channel membrane geometry.

Beyond the constructive issues concisely addressed above, altogether difficult hindrances to be overcome are to be expected even on the theoretical modeling of

Fig. 6.19 Example of a branched channel membrane produced with the iCVD technique. Reproduced from [84], with permission

microscale ECMOs. In this regard, the most peculiar aspect associated with scaling down the device is given by the fact that the $10\,\mu m$ limit sought for the characteristic dimension of the channel cross-section is comparable to the size of red blood cells, an issue that brings into play new phenomena both for the hydrodynamics and the mass transport processes that take place in the exchanger. Specifically, as regards the blood microhydrodynamics, the rheological behavior of blood is yet to be completely understood, and constitutive relationships such as the Bullik power law model [85] or other approaches [86] used to characterize flow regimes in hollow fiber exchangers must yet be validated at these length scales. In turn, 3 different blood rheologies as well as modified flow regimes should also be expected to have a significant impact on mass transport in the blood phase, and therefore, the validity of model predictions on overall mass transfer coefficients tested in hollow fiber modules for both oxygen [86] and carbon dioxide [66] should not be taken for granted. Because in microchannels of $10\text{-}\mu m$ cross-sectional dimension, red blood cells can be expected to flow "one at a time", it appears sensible to assume that the oxygen uptake and carbon dioxide discharge will be described by a time-periodic model, where the frequency with which a given portion of the membrane surface is visited upon by the streaming red blood cells introduces a new timescale that is altogether absent in larger hollow fiber modules. These observations make it clear that the rational design of microfluidic-assisted ECMOs passing the benchmark test of clinical practice will be the result of the synergetic cooperation between researchers of many different branches of science, experimentalists, and theoreticians alike.

Acronyms

AV   Arteriovenous
$AVCO_2R$ Arteriovenous carbon dioxide removal

ARDS	Acute respiratory distress syndrome
CPB	Cardiopulmonary bypass
DPG	Diphosphoglycerate
ECCO2R	Extracorporeal CO_2 removal
ECMO	Extracorporeal membrane oxygenator
ELSO	Extracorporeal Life Support Organization
EL	Extraluminal (flow)
iCVD	Initiated chemical vapor deposition
IL	Intraluminal (flow)
MV	Mechanical ventilation
PEEP	Positive end-expiratory pressure
PMEA	Poly(2-methoxyethylacrylate)
PMP	Polymethylpentene
PP	Polypropylene
VV	Veno-venous

Symbols

A	Total exchange area
a	Specific exchange area per unit volume
c	Concentration in blood
d	Capillary diameter
d_f	Fiber
d_p	Pore diameter
$\mathcal{D}$	Diffusivity
E	Enhancement factor
F_B	Blood volumetric flow rate
Gz	Graetz number
H	Henry's constant (liquid composition as molar fraction)
H'	Henry's constant (liquid composition as molar concentration)
Htc	Hematocrit
$\mathcal{H}$	Apparent Henry's constant
$K_1, \ldots, K_4$	Oxygen–heme-binding constants
K_{a1}, K_{a2}	First and second carbonic acid dissociation constants
K_c	Overall mass transfer coefficient (concentration driving force)
K_h	Carbon dioxide hydration equilibrium constant
K_p	Overall mass transfer coefficient (pressure driving force)
k_p	Gas-phase mass transfer coefficient (pressure driving force)
$\mathcal{K}$	Constant defined in Eq. 6.7
L	Length (capillary or channel)
ℓ	Characteristic length
N_{tm}	Transmembrane gas flux
Q_B	Blood flow rate
Pe_{tm}	Transmembrane Peclet number
Pr	Prandtl number
p_{50}	Oxygen partial pressure at 50 % of hemoglobin saturation

p	Partial pressure
$\mathscr{P}$	Membrane permeance
R	Dimensionless parameter defined in Eq. 6.56
R_0	Inner radius
$\mathbb{R}$	Gas constant
Re	Reynolds number
$S_\%$	Oxygen fractional saturation
T	Temperature
V_i	Volumetric flow rate of inspired air
Y	Dimensionless parameter defined in Eq. 6.56
v	Blood velocity
α_B	Gas partition coefficient in blood
α_m	Gas partition coefficient in membrane
δ	Thickness
ε_f	Void fraction of the fiber bundle
ε_w	Membrane wall porosity
θ	Contact angle
σ	Gas–blood surface tension
τ	Tortuosity factor

Subscripts

a	Arterial blood
alv	Alveolar air
b	Chemically bound form of gas dissolved in blood
CO_2	Carbon dioxide
Hb	Hemoglobin
i	Inspired air
in	Inlet
m	Molecular (free) form of gas dissolved in blood
O_2	Oxygen
out	Outlet
RBC	Red blood cell
v	Venous blood

Superscripts

*	In equilibrium conditions
0	In absence of chemical reactions
B	In blood
G	In gas phase
i	At the interface
m	In the membrane
p	In the pores of the membrane

Overscripts

$\sim$ Dimensionless

Chapter 7
Artificial Kidney

7.1 Introduction

The introduction of artificial kidneys as devices able to replace the function of healthy kidneys marks the beginning of the history of artificial organs. Nowadays, artificial kidneys are the most widely used artificial organs in clinical practice and have become almost a synonym of dialysis, since they are always implemented as membrane-based separation devices used to remove toxins and the excess water from patient blood. More than 1.7 million people in the world suffering from final-stage renal diseases, in which the renal function is seriously or completely compromised, rely on some form of hemodialysis to keep them alive. Despite the wide diffusion of artificial kidneys and the progress achieved in terms of survival and compliance of the patients, dialysis systems—in all the forms in which they are implemented—are far from being considered as true artificial kidneys: in fact, they only replace the detoxification function of healthy kidneys and only do so poorly. Efforts are being carried out to develop more patient-friendly hemodialysis systems or to develop bio-artificial devices able to best simulate the function of kidneys.

This chapter is mainly focused on the engineering aspects of the design of the dialysis systems, on the basis of the fundamentals already discussed in Chap. 3. After a short overview of the functions and pathologies of the renal system aimed at defining the medical requirements for the artificial devices, we report a summary of the historical development of the artificial kidney. Subsequently, the different ways in which membrane separation is implemented in renal replacement therapies, as well as the characteristics of membrane modules used, are reported. On the basis of this information, mathematical models are then developed to describe the behavior of these systems and optimize their design. Patient-device models are also considered to predict the evolution in time of toxin levels in patient blood and optimized the dialysis protocol. The chapter ends with a survey on the latest technological developments in terms of portable or wearable devices as well as on bio-artificial devices aimed at replacing the functions of natural kidneys.

© Springer-Verlag London 2017
M.C. Annesini et al., *Artificial Organ Engineering*,
DOI 10.1007/978-1-4471-6443-2_7

Fig. 7.1 Nephron structure and scheme of the processes occurring in nephrons

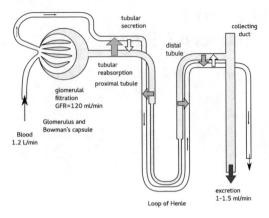

7.2 Structure and Function of Renal System

The kidneys and renal apparatus perform several functions, namely removal of the end products of body metabolism, elimination of excess water, control of the electrolyte homeostasis, and regulation of blood pressure. The kidneys are paired solid organs, located in the posterior abdomen, which process the blood supplied by the renal arteries and, after detoxification, return it to the circulatory system via the renal veins; the waste fluid produced in this process is collected into the ureters and, eventually, into the urinary bladder. Each kidney is made up of 1–1.5 millions of mass transfer units called *nephrons*, working in parallel; each nephron is made up of two subunits in series (see Fig. 7.1): the *glomerulus,* which is designed to ultrafiltrate the blood and remove the metabolic products, and the tubules, which provide fine regulation of the excretion of water and solutes in the urine. In addition to the basic excretory and regulatory functions, the kidneys provide a critical endocrine function, producing erythropoietin, vitamin D, angiotensin, prostaglandins, and other endocrine compounds.[1]

Blood is fed to the kidney through the renal arteries at a rate of about 1.2 l/min (about 20 % of the cardiac output) and distributed to the nephrons, where about 20 % of the plasma (120 ml/min) is forced through the glomerular membrane into the Bowman's capsule. The difference in the hydrostatic pressure and osmotic/oncotic pressure between the glomerular capillaries and the Bowman's capsule is the driving force for ultrafiltration through the globular membrane; glomerular ultrafiltration rate (GFR) is then given by:

$$\text{GFR} = K_{GF} \left(\Delta P - \Delta \pi \right) \tag{7.1}$$

---

[1] This and the following section were authored by Sergio Morini—Center for Integrated Biomedical Research (CIR), University "Campus Bio-Medico" of Rome, Italy.

Table 7.1 Glomerular filtration of different solutes

Solute	Molecular weight	Sieving
Water	18	1
Glucose	180	1
Insulin	5000	1
Myoglobin	17000	0.75
Hemoglobin	68000	0.01
Albumin	69000	0.0001

Sieving is defined as the ratio of filtrate to plasma concentration [87]

where K_{GF} is the hydraulic permeability, ΔP the difference in the hydrostatic pressure between the blood capillaries (about 55 mmHg) and the filtrate (about 15 mmHg), and $\Delta\pi$ the difference in the osmotic/oncotic pressure; since the filtrate is virtually free of proteins, $\Delta\pi$ is equal to the colloid osmotic pressure of plasma proteins. It is worth noting that the capillary hydrostatic pressure in glomerular capillaries is higher than that in other capillaries and pressure drop along the capillary length is very low; therefore, despite the significant increase in the oncotic pressure due to the large volume of filtrate (from 20 to 30 mmHg), outward filtration of plasma occurs all along the capillary. Indeed, an autoregulation mechanism—based on capillary constriction or dilatation—acts to maintain a constant renal blood flow rate and glomerular ultrafiltration rate even if the arterial blood pressure varies in a quite wide range (80–170 mmHg) [87].

The glomerular membrane acts as a barrier for the blood cells and almost all plasma proteins, while electrolytes and small solutes (glucose, amino acids, urea, creatinine, etc.) flow in the glomerular filtrate at the same concentration as in plasma. The selectivity of the membrane is based on the solute molecular size, so that molecules with molecular weight (MW) lower than 10000 kDa are freely filterable; larger molecules experience a progressively higher restriction to passage in the filtrate, and molecules larger than 100000 kDa are completely rejected (see Table 7.1). The net electrical charge also plays a significant role in the filterability of molecules, due to the negative charge present in the glomerular membrane; therefore, albumin, which is a negatively charged macromolecule at physiological pH, is almost completely retained.

The glomerulus provides a filtration rate of about 120 ml/min—with wide physiological fluctuations during the day—so that the plasma is wholly filtrated about 60 times a day; such a high GFR is due to the high-filtration area (about 2 m^2) and to the unusually high hydraulic permeability of the glomerular capillary, which is two orders of magnitude higher than for most of the capillaries in the human body. Normal GFR corresponds to a highly redundant physiological system, since the excretion of a 1–2 l volume per day would be sufficient to clear the metabolic waste products. Actually, the glomerular filtrate flows downstream through the system of renal tubules (the proximal tubule, the loop of Henle, the distal tubule, and collecting duct) that provides for the reabsorption of most part of the water (more than 99 %)

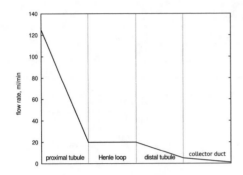

	Reabsorption
water	>95%
glucose amino-acids	~100%
ions $(Na^+, K^+, Ca^{2+}, Cl^-, HCO_3^-)$	~99.5%
uric acid	~91%
urea	~50%
creatinine	0%

Fig. 7.2 Water flow rate profile in the tubular reabsorption process

and solutes filtered in the glomerulus (see Fig. 7.2), as well as to the secretion of other solutes, including endogenous metabolites (e.g., creatinine, uric acid) and drugs (e.g., penicillin, furosemide, cisplatin). While glomerular filtration is quite an unselective process, transport in the tubular system is very selective and controls the amount of different substances excreted with the urine. Here, the driving forces for water or solute transport are mainly osmotic and oncotic pressure differences, but specialized cells provide the active transport mechanisms based on the Na^+/K^+ pump to create the concentration gradient and control the mass flow of selected components. Some points are worth being remarked:

- glucose, which crosses through the glomerular membrane in the Bowman's capsule, is completely reabsorbed into the renal tubules, unless its blood concentration exceeds a critical value; therefore, there is no glucose in the urine of healthy subjects;
- creatinine (an end product of protein metabolism) is removed from plasma in the glomerular filtration and it is not reabsorbed in the tubular system. Indeed, creatinine concentration in the urine is a measure of the glomerular filtration rate.[2]
- urea permeates through the glomerular membrane, but about 50 % of urea is reabsorbed in the proximal tubule; in the distal tubule, urea is secreted and reabsorbed with a rate depending on the water diuresis.
- uric acid, as well as few other organic ions, is both reabsorbed and secreted; actually, reabsorption dominates and only 10 % of the filtered uric acid is excreted; drugs that inhibit tubular reabsorption can be useful to improve uric acid removal.
- concentrations of different solutes vary along the tubule, depending on the reabsorption and secretion process and water reabsorption: while concentrations of some substances such as glucose and amino acids fall along the tubules, other substances that are reabsorbed to a lesser extent or not at all concentrate along the tubule as a result of water reabsorption.

---

[2]A small secretion of creatinine in proximal tubule occurs, and this affects the evaluation of the GFR.

7.3 Renal Diseases and Medical Requirements for a Renal Replacement Therapy

Renal diseases can arise from prerenal (inadequate blood flow), intrarenal (nephron damage: glomerulopathy and tubulointerstitial diseases), or postrenal (obstruction, structural defects) problems. Such diseases can result in:

- *acute kidney injury* or *acute renal failure* (AKI or ARF), characterized by rapid deterioration of renal functions—mainly a rapid decrease of GFR—and retention of toxins, fluids, and end products of metabolism. Usually, ARF is reversible, but incomplete recovery can lead to an insufficient renal function, evolve to chronic renal disease, or progress to end-stage renal disease. Medical treatments are based on dietary restrictions, pharmacological therapy aimed at treating the underlying causes and improving diuresis, or *renal replacement therapy* (RRT), including hemodialysis, peritoneal dialysis, and *continuous renal replacement therapy* (CRRT); in particular, RRT may be necessary to cover the renal function during the time required for treating the renal pathology.
- *chronic kidney disease* (CKD), resulting from gradual, progressive loss of renal function. It is usually caused by chronic glomerulonephritis, ascending infection of the urinary tract, hypertension, and vascular disease; in some cases, it results from progression of acute renal failure. Since in normal physiological conditions the kidney system is highly redundant, clinical symptoms appear only when more than 75 % of the renal function is lost and a replacement therapy is required when the functional loss exceeds 90–95%. Indeed, international guidelines define CKD as sustained kidney damage indicated by the presence of structural or functional abnormalities (e.g., microalbuminuria/proteinuria, hematuria, histological or imaging abnormalities) and/or reduced glomerular filtration rate (GFR) to less than 60 ml/min per 1.73 m^2 body surface area for at least 3 months. Five levels of chronic kidney disease are identified (see Table 7.2), depending on the GFR values: while in the first stages the therapeutic intervention, including dietary restriction, is aimed at controlling comorbidity and slowing down the disease progression, treatment with dialysis or renal transplantation is the only option in the fifth stage of CKD (end-stage renal disease, ESRD).

Table 7.2 Stages of chronic kidney disease [88]

Stage	Description	GFR, ml/min[a]
1	Kidney damage with normal or increased GFR	>90
2	Kidney damage with mild decreased GFR	60–89
3	Moderately decreased GFR	30–59
4	Severely decreased GFR	15–29
5	Kidney failure (ESRD)	<15

[a]Referred to 1.73 m^2 of body surface area

When the kidneys cannot remove the metabolic wastes, substances normally eliminated with urine accumulate in blood and other body fluids and result in an overall deterioration of the biochemical and physiological functions; plasma concentrations of creatinine and urea (which are highly dependent on glomerular filtration) begin a hyperbolic rise as GFR diminishes. It appears that 8 ml/min is the lowest level of creatinine clearance compatible with survival; urea nitrogen and creatinine levels higher than 80–100 mg/dl and 5–15 mg/dl require RRT. Indeed, urea and creatinine are not the most important contributors to the uremic symptoms and they become toxic at concentrations higher than those encountered in uremia; however, urea and creatinine are used as markers to detect impaired renal elimination and body accumulation of many other substances usually referred to as uremic toxins. There is a general agreement that the uremic syndrome is the result of the overall accumulation of multiple interfering substances [89, 90], usually classified depending on their molecular weight or their ability to bind to proteins and then to behave as larger molecules (see Table 7.3).

Therefore, severe uremia caused by acute or chronic renal failure requires a RRT providing for the removal of a variety of uremic toxins. These therapies include extracorporeal dialysis (with its variations as hemodialysis, hemofiltration, or hemodiafiltration as discussed in the following section) or intracorporeal peritoneal dialysis. Extracorporeal dialysis requires blood circulation through an external artificial device that relies on a selective membrane for blood detoxification; intermittent short dialysis sessions (about 10 h, divided in three sessions per week) with a high removal rate of uremic toxins are generally used in chronic renal disease, while continuous hemodialysis (or CRRT) with low blood flow and solute removal rate can be used

Table 7.3 Examples of uremic toxins and their classification

Low MW		Middle and large MW		Protein-bound compounds	
Compound	MW	Compound	MW	Compound	MW
Urea	60	Prostaglandins	600	Homocysteine	135.2
Phosphate	80	Parathormone (PTH)	9424	Hippuric acid	1729.2
Creatinine	113	Atrial natriuretic factor	3080	Indoxyl sulfate	213.2
Histamine	127	β−endorphin	3465	Phenols and indols	
Catecholamines	160	Endothelin	4238	Polyamines	
Uric acid	168	Adrenomedullin	6032		
		β_2-microglobulin	11818		
		Leptin	16000		
		Cytokines	15000–30000		
		Immunoglobulin LC	28000–56000		
		Myoglobin	16700		

mostly in the intensive care units to face acute renal failure. Obviously, an intermittent treatment is more compatible with an acceptable life standard, but results in a sawtooth time course for solute concentrations, which can also lead to hemodynamic stability issues; on the contrary, continuous treatments allow to maintain a more stable solute concentration and may be suggested for AKD associated with hemodynamic instability.

Peritoneal dialysis exploits the properties of the peritoneal membrane to exchange water and solutes with a washing solution; in detail, dialysate is infused in the peritoneal cavity, left for toxin and water transfer and then drained; the procedure is repeated several times, using a new washing solution. The process can be automated with a device programmed to deliver a predetermined volume of dialysate and to drain the peritoneal cavity at fixed intervals. Peritoneal dialysis is not covered in this book, since it does not involve the use of artificial mass transfer devices.

Clearly, an ideal RRT should reproduce the physiological processes carried out by healthy kidneys. It is evident that artificial devices do not allow to replace the endocrine renal functions, including synthesis of erythropoietin and vitamin D, for which a substitutive pharmacological therapy must be provided; however, also with regard to blood detoxification, the conventional dialytic treatments are currently far from adequately replacing the kidney function. Even if dialysis can provide the conditions for the survival of the patient, it results in poor quality of life and high morbidity and mortality. Progresses in RRTs have been obtained by improving the membrane properties (high-flux membranes), exploiting different mass transport mechanisms (convection versus diffusion) and modifying scheduling and duration in the intermittent treatments. Nevertheless, considerations about the complex control of renal excretion, based on an almost unselective filtration followed by a system of reabsorption and secretion processes carried out by highly specialized cells, suggest that a simple process based on the selectivity of a synthetic membrane cannot adequately mimic the functions of the renal system. Unlike the healthy kidney, a dialyzer does not include a mechanism to reabsorb the permeated water and solutes; therefore, an artificial kidney must be designed and its operating conditions chosen so as to limit the actual removal of excess water or solutes; otherwise, re-infusion of a replacement fluid may be required. Usually, dialysis is efficient for removing low MW solutes, such as urea and creatinine; insufficient removal of middle MW molecules is often suggested as a cause of the poor long-term outcome of hemodialysis. Furthermore, the unphysiological pattern of intermittent dialysis with a fast water removal and change in toxin concentration, followed by a prolonged water and toxin accumulation period, results in numerous intra- and interdialytic clinical problems.

7.4 Historical Development of Artificial Kidney

The idea to use dialysis in the clinical practice dates back to the middle of nineteenth century, when the "father of dialysis," Thomas Graham, predicted that some of his findings on the removal of solutes from fluids containing colloids and crystalloids might be applied in medicine. Actually, it was in 1913 that Abel and coworkers, in Baltimore, attempted to use a "vividiffusion" device—for which they coined the name of "artificial kidney"—to treat the blood of dogs in extracorporeal circulation [91]. Abel's dialyzer consisted in a series of tubes of semipermeable collodion membrane with a diameter of 8 and 40 cm long; the membrane surface area was about $0.32\,m^2$, too small for use in human patients. Few years later, starting from 1924, the German physician George Hass performed the first dialysis tests in humans: he used again collodion tubes with a larger membrane area ($1.5–2\,m^2$) in dialysis sessions lasting from 15 to 60 min. He proved that the procedure could be well tolerated, even if such a short treatment had no significant therapeutic effects [92].

Two main problems had to be solved to bring the artificial kidney to the clinical use: to find a nontoxic and reliable anticoagulant and to be able to produce large, defect-free semipermeable membranes. It was at the end of 1930s that purified heparin became readily available for human application and a new material, known as cellophane, was marked for commercial use; heparin had less adverse effects than hirudin, used in the first dialysis trials, and cellophane seemed to be an excellent material for dialysis membranes.

The cooperation of a young physician, Willem Kolff, and an engineer, Hendrik Berk, resulted in the construction of the first dialyzer suitable for human application in the Netherlands. Kolff himself described how, in 1939, he was pushed to think to an apparatus that could replace the renal function after the miserable death of a 22-year-old uremic patient, in the Groningen University Hospital: Kolff's idea was that if he could have removed 20 g of urea per day from his blood, the young man might have survived [43]. In the subsequent years, Kolff tested the properties of cellophane membranes and evaluated the membrane area required for an effective dialysis, until he moved to Kampen, after the Nazi occupation of the Netherlands. Here, Kolff and Berk constructed a dialyzer with a large membrane area. The device consisted of a horizontal cylindrical rotating drum; a 30 or 40 m long cellophane sausage tubing, 2.5 cm wide, was wound on the cylinder and perfused with the patient's blood; the lower half of the rotating drum was dipped in a bath of 70 l of dialysis fluid so that as the cellophane tube passed through the bath, the toxins would pass from the blood to the rinsing solution. Between 1943 and 1945, Kolff tried unsuccessfully to treat 16 patients with his rotating drum kidney, until the first therapeutic success arrived in 1945 with a 67-year-old woman, a Nazi collaborator, who was dialyzed for 11 h with a dramatic improvement of her conditions; diuresis started within one week, and the patient was released from the hospital with a normal renal function.

After the end of the World War II, the Kolff machines were sent and used in several hospitals in Europe and in USA and other devices were built with minor or major modifications and improvements. In 1947, Nils Alwall in Lund (Sweden) designed the first dialyzer with controlled ultrafiltration, which was able to remove the excess water from the patient body: in this device, the cellophane membrane, which was wrapped around a stationary vertical drum, was supported by two protective metal screens; negative pressure was applied to the rinsing fluid to allow ultrafiltration. In the years from 1947 to 1959, two new types of artificial kidney were realized: the stationary *coil-type* artificial kidney, with the cellophane membrane wrapped with a separator that allowed the compact device to be perfused by the dialysate, and the first *parallel flow dialyzer*, with sheets of membranes separated by rubber pad, with both blood and dialysate flowing parallel to the membrane sheets in opposite directions. In 1955, the first *twin-coil dialyzer*, a quite compact device with a membrane area of about $2\,m^2$ and priming volume of 750 ml, was realized by Kolff: with this device, an urea clearance of 140 ml/min and significant ultrafiltration were obtained.

In these early years, dialysis was limited to the treatment of acute and reversible renal failure: actually, the use of dialysis in the treatment of chronic patients was not limited by the performance of the devices, but by the difficulties in gaining repeated access to the patient's circulation without compromising the blood vessels. The first solution became available in 1960, with a semipermanent arteriovenous (AV) cannulation (Scribner shunt) that allowed to perform repeated treatments, once or more weekly, on patients with chronic terminal renal failure; further improvement was obtained in 1966 by Brescia, Cimino, and coworkers, with the surgically created AV fistula, which did not require exteriorised shunts [93]. With the introduction of these procedures, the treatment of chronic uremic patients by intermittent hemodialysis became a standard, widely used therapeutic method.

Starting from all these pioneering efforts, technological research has been devoted to increase the efficiency and safety of dialysis, as well as to improve the patient compliance in chronic treatments, miniaturize the equipment, and reduce the costs. A large number of dialyzers were commercialized, both disposable and non-disposable. The performance of the coil dialyzer was improved with the use of thin cuprophane membranes, while in 1965 hollow fiber modules with a high membrane area-to-volume ratio and efficient mass transfer kinetics were introduced. Early commercial hollow fiber devices consisted of a bundle of fibers packed in a Perspex® (polymethyl methacrylate) tubular housing, with a configuration similar to a shell and tube heat exchanger. The bundle was bonded at the two ends, and a screw cap manifold was used. Blood entered and left the device via the manifolds, which were designed to optimize both blood velocity and pressure drop, thereby ensuring an even distribution of blood in the fiber bundle. The advantages of hollow fiber dialyzers were apparent: improved blood flow path through the capillary membranes led to a low boundary layer resistance and enhanced mass transport; furthermore, also the production costs were lower. Thereafter, capillary fiber modules became universally used for long-term dialysis.

Further significant improvements have been obtained from the 1960s until now; among these:

- the *single-pass* dialysis and the introduction of proportioning pumps to dilute the dialysate solutions in central systems for simultaneous supply of several artificial kidneys
- the development of new synthetic membranes, more permeable and biocompatible than those based on cellulose derivatives

- the introduction of high-flux, highly permeable membranes to improve the removal of medium molecular weight compounds, such as β_2-microglobulins
- the clinical use of hemofiltration, with removal of solutes by convection through highly permeable membranes, as a different procedure to replace the renal function.

7.5 Membrane-Based Processes for Renal Replacement Therapies

While the term "dialysis" is often used in a generic sense as a synonym for RRT, it is important to recognize that there is a variety of membrane-based processes that are applied in the clinical practice both for acute and for chronic renal failures.

Indeed, as discussed in Chap. 3, different membrane processes can be considered, depending on the driving force that determines the mass flow through the membrane: the dialysis process is characterized by the diffusive solute transport through the membrane, which is driven by the solute concentration gradient; differently, in the ultrafiltration process, the mass transfer through the membrane is mainly due to solvent flux driven by the transmembrane pressure difference; solutes also cross the membrane driven by convection, i.e., dragged by the solvent flux. In clinical practice, both diffusion-based and convection-based separation processes are used in RRTs.

7.5.1 Hemodialysis (HD)

Hemodialysis (see Fig. 7.3) is a membrane-based process in which the toxin removal occurs mainly by toxin diffusion through the membrane; in this case, the driving force for mass transfer is given by the concentration difference between blood and a rinsing solution, the *dialysate*, placed downstream of the membrane. The dialysate is obtained by diluting a concentrated solution of electrolytes and, in some cases, glucose; more specifically, such solution contains sodium, chloride, and magnesium ions at the same concentration as in normal plasma (monitoring and adjusting of the potassium level is required to control the serum potassium concentration); furthermore, acetate or bicarbonate is added to buffer the solution pH. A fine-tuning of the dialysate composition is prescribed to adapt the therapy to the patient's need.

In a pure dialysis process, with no solvent flux through the membrane, the solute flux is given by

$$J_s = \frac{\mathscr{P}}{\delta} (c_1 - c_2) \tag{7.2}$$

where $\mathscr{P}$ is the membrane permeability, δ the membrane thickness, and the driving force term, $c_1 - c_2$, is the difference of the solute concentration between the liquid phases at the two sides of the membrane. The solution–diffusion model suggests that $\mathscr{P}$ depends on the solute diffusion coefficient in the membrane and on its partition

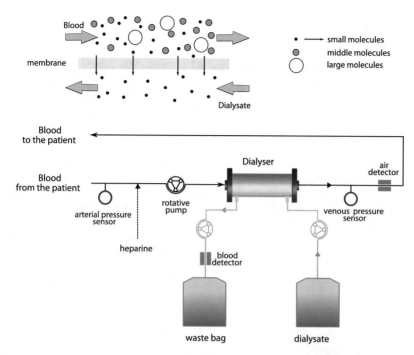

Fig. 7.3 Hemodialysis process. On the *top*: transport of small MW solute through the membrane occuring by diffusion; on the *bottom*: schema of a hemodialysis circuit

coefficient between the membrane and the liquid phase (see Chap. 3). Actually, $\mathscr{P}$ is usually high for low MW solutes such as urea (60 Da) and creatinine (113 Da), but decreases with increasing MW; as a result, hemodialysis is effective for the removal of small toxins, while toxins with MW exceeding 500 kDa are not significantly eliminated.

In clinical hemodialysis, typically 200–400 ml/min of blood is drawn to the dialyzer, so that the whole patient's blood is treated every 15 min; the ratio of dialysate to blood flow rate is usually set in the range 1–1.5. The pressure difference between blood and dialysate stream (100–500 mmHg) results in some water flux through the membrane and helps to remove the excess water in the body (2–4 l during a session). When high-flux membranes such as polyacrylonitrile or polysulfone membranes are used, the convective transport of middle MW molecules can occur even in hemodialysis; the convective transport in this case may be relevant to improve the clearance of middle MW molecules.

7.5.2 Hemofiltration (HF)

Hemofiltration (see Fig. 7.4) is essentially an ultrafiltration process, in which mass transfer across the membrane is mainly due to the solvent flux through a high-flux membrane with large pores; the driving force is given by the transmembrane pressure difference. Molecules that are small enough to pass through the membrane pores are dragged across the membrane by convection. The water removal rate is, in this case, largely higher than that required to remove the excess water in the body, and the volume balance is maintained via the administration of a substitution fluid that must be sterile, non-pyrogenic, and endotoxin-free. Infusion of the substitution fluid may be carried out either downstream (postdilution) or upstream (predilution) of the membrane module. The latter option allows to achieve very high ultrafiltration rates, with a higher clearance of middle MW molecules.

In hemofiltration, the solute removal rate is then proportional to the ultrafiltration rate, which is greater than that in the case of hemodialysis; as discussed in detail in Chap. 3, the solute flux, J_s, is given by

$$J_s = \bar{C}_s (1 - \sigma) J_v \tag{7.3}$$

where $\bar{C}_s$ is the mean solute concentration in the membrane and J_v the volume flux through the membrane; σ is the Staverman reflection coefficient that varies from 0—for solutes that are small compared to the diameter of the membrane pores and can cross the membrane unhindered—to 1—for solutes larger than membrane pores, which are completely rejected.[3]

With the high-flux membranes used in hemofiltration, high removal rates can be obtained for solutes with MW up to several thousands of daltons, while macromolecules such as proteins or cellular elements are retained.

7.5.3 Hemodiafiltration (HDF)

In hemodialysis and hemofiltration, the solute transport is driven primarily by diffusion and convection, respectively. Between these two limiting conditions, a variety of intermediate treatment modalities—generally referred to as hemodiafiltration—have been proposed or are in clinical use; these processes basically consist in hemodialysis combined with controlled ultrafiltration; the ultrafiltration rate is higher than that in hemodialysis but lower than that in hemofiltration (see Fig. 7.5). Hemodiafiltration processes aim at combining the best features of hemodialysis and hemofiltration, i.e., the high clearance of small solutes obtained by diffusion and the high clearance of middle MW molecules obtained by convection.

---

[3]In medical literature, the *sieving coefficient* $\mathscr{S} = 1 - \sigma$ is also used. Therefore, for an unhindered solute $\mathscr{S} = 1$, while for a component completely rejected $\mathscr{S} = 0$.

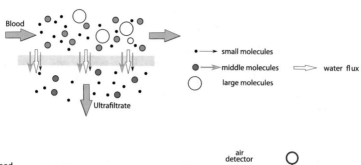

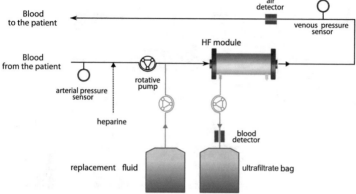

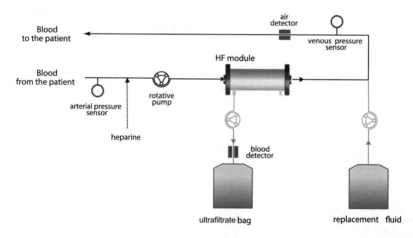

Fig. 7.4 Hemofiltration process. On the *top*: large solvent flux through a porous membrane drags small and middle MW molecules by convection; on the *bottom*: scheme of a hemofiltration circuit with pre- or postdilution

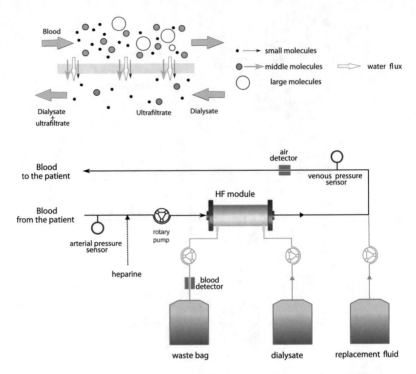

Fig. 7.5 Hemodiafiltration process. On the *top* solute transport by convection and diffusion; on the *bottom* schema of a hemodiafiltration circuit with postdilution

A comparison of the removal efficiency as a function of the molecule size in conventional hemodialysis, high-flux hemodialysis, and hemofiltration is reported in Fig. 7.6.

Fig. 7.6 Comparison of profiles of solute clearance obtained in conventional dialysis (HD), high-flux hemodialysis (HD/high-flux, low ultrafiltration rate), hemofiltration (HF, with high ultrafiltration rate and postdilution), or hemodiafiltration (HDF). In the plot, the size of urea, β_2-microglobulin (β_2M), and albumin are reported as a reference

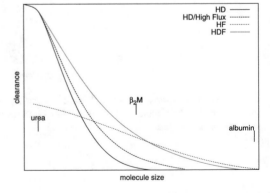

7.5.4 Continuous Renal Replacement Therapy (CRRT)

While in chronic renal diseases the dialytic treatment is usually carried out intermittently, typically for 3–4 h three times a week, CRRT is an extracorporeal blood detoxification process that is applied 24 h/day to replace the impaired renal function, mainly with critically ill patients in intensive care units. Different modalities of treatment can be implemented, depending on the blood access (arteriovenous without pump or venous-venous with pump) and the method of solute removal (convection in ultrafiltration or hemofiltration, diffusion in hemodialysis, convection and diffusion in hemodiafiltration): slow continuous ultrafiltration, to remove the excess water, continuous arteriovenous or venous-venous hemofiltration, continuous arteriovenous or venous-venous hemodialysis at low blood and dialysate flow rate, and continuous arteriovenous or venous-venous hemodiafiltration. A summary of the operating conditions is summarized in Table 7.4 [94]. These techniques are similar in principle to those previously described for intermittent therapies, but are continuously applied with smaller flow rates; they provide a smooth removal of solutes and water, thus reducing hemodynamic stability issues.

7.6 Membrane Modules for Hemodialysis

Different types of polymeric membranes are now used for renal replacement therapies. Two main classes of materials can be considered: unsubstituted (regenerated) or modified cellulose and synthetic membranes [95–97] (see Table 7.5).

In the early days of clinical use of hemodialysis, cellulosic membranes were used. Cellulosic materials are constituted by monomeric subunits of cellobiose, a saccharide with a ring structure rich in hydroxyl groups. These polymers are therefore extremely hydrophilic and sorb water, forming a macroscopically homogeneous hydrogel; due to the crystalline structure of cellulose, very thin membranes (5–15 μm) can be formed with good efficiency for the removal of low MW molecules, but poor

Table 7.4 Conditions adopted for continuous renal replacement therapies [94]

	Access	Q_B, ml/min	Q_D, ml/min	Q_F, ml/min	Re-infusion
Slow ultrafiltration	AV	50–100	–	2–6	No
	V-V	50–200	–	2–8	No
Hemofiltration	AV	50–100	–	8–12	Yes
	V-V	50–200	–	10–20	Yes
Hemodialysis	AV	50–100	10–20	1–3	No
	V-V	50–100	10–30	1–5	No
Hemodiafiltration	AV	50–100	10–20	8–12	Yes
	V-V	5100–200	20–40	10–20	Yes

Table 7.5 Membranes for hemodialysis

Cellulosic membranes		Synthetic membranes
Unmodified	Modified	
Cuprophane	Cellulose acetate	Polysulfone (Biosulfane)
Cellophane	Cellulose diacetate	Polyamide
	Cellulose triacetate	Polyacrilonitrile (AN69, SPAN, PAN)
	Cellulose 2–5 diacetate (Diaphan)	polymethylmethacrylate (PMMA)
	Hemophan (tertiary amine)	Polyethylvinylalcohol (EVAL)
	Synthetically modified cellulose (benzyl group)	Polycarbonate (Gambrane)
	Saponified cellulose ester (SCE)	

permeability for middle MW molecules. The main problem with these membranes is related to the presence of free hydroxyl groups that are responsible for the complement activation and leukopenia. In order to solve this problem, modified cellulosic membranes with a fraction of hydroxyl groups replaced with other moieties have been manufactured; usually, acetate groups substitute approximately 70–80 % of the hydroxyl groups in acetate, diacetate, and triacetate cellulose. The substitution proved to be useful to mitigate the complement activation; furthermore, modification results in some increase of the pore size, yielding a higher water permeability and higher clearance for middle MW molecules. Substitution of a small fraction of the hydroxyl groups with bulky groups, such as tertiary amines (in commercial Hemophan®) or benzyl groups, which sterically reduce the interaction with complement activation products, also proved effective to improve the membrane biocompatibility. Actually, the technological developments allowed to obtain biocompatible cellulosic membranes with a wide spectrum of physicochemical properties.

However, it was evident that, even with modified cellulosic membranes, residual hydroxyl groups still caused high complement activation; therefore, a progressive transition toward synthetic membranes started from the 1970s. Different polymers, such as polyacrylonitrile (e.g., AN69), polymethylmethacrylate (PMMA), polysulfone, and polyamide, with different monomeric subunits have been introduced for clinical use; these membranes are primarily made of hydrophobic polymers (with hydrophilic additives) without hydroxyl groups and offer excellent biocompatibility. The large pore size and the higher hydraulic permeability make these membranes very interesting for high-flux hemodialysis or for hemofiltration; in particular, synthetic membranes can be manufactured with different pore sizes with high clearance also for larger molecules.

An higher thickness (>20 µm) enhances the mechanical resistance of the membrane and the possibility to withstand high ultrafiltration rates, but reduces the permeability of smaller molecules. The introduction of asymmetric membranes with a thin (<1 µm) selective skin supported on a spongy layer that provides the mechanical

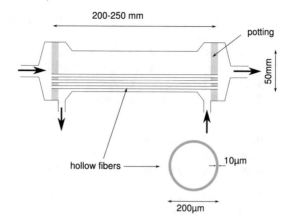

Fig. 7.7 Scheme of a hollow fiber membrane module for hemodialysis

resistance allows now to independently control the diffusive and convective transport properties.

Nowadays, membranes used in the clinical practice are hollow fiber membranes assembled in modules (see Fig. 7.7), with the fiber bundle enclosed in a dialyzer housing; to build the module and avoid the contact between blood and dialysate, the membrane fibers are potted with a polymeric compound (usually polyurethane). To fulfill the medical requirements, dialyzers of different sizes are commercialized, with membrane areas spanning from 0.2 up to about $2\,m^2$, with up to 15000 fibers or more, and a priming volume of 50–150 ml.

The proper choice of the materials for all the components, the design of the whole fiber bundle, and the design of blood and dialysate flow patterns are fundamental to ensure a good performance of the membrane module. It is evident that all the materials for membrane housing and membrane potting must be suitable for biomedical use, should not release any components in the fluid, and have to withstand the sterilization procedures. Furthermore, at least for small molecules, the resistances to mass transport in blood and dialysate, which are affected by the thickness of blood and dialysate boundary layers near the membrane surface, are usually of the same order of magnitude of the membrane resistance, and the fluid flow pattern plays a fundamental role in the device performance; an uneven distribution of blood and dialysate, with channeling and formation of preferential pathways, must be avoided.

Most hollow fibers have an inner diameter of 180–220 μm and a length of 20–25 cm. It is worth noting that the smaller the fiber diameter, the lower the blood-side mass transport resistance, but also the higher the flow resistance and pressure drop in the module.

In Chap. 3, we have shown that the transport properties of a membrane can be described in terms of three phenomenological parameters:

- the diffusive permeation coefficient of the solute, ω—or the membrane permeance—that represents the ability of the solute to diffuse through the membrane (i.e., the ratio of the solute flux to the difference in solute concentrations in the case of no volume flux)

- the hydraulic permeability, L_p, that, in a pure solvent permeation test, is given by the ratio of the volumetric flux through the membrane to the transmembrane pressure difference
- the reflection coefficient (Staverman coefficient) σ that measures the ability of the solute to cross the membrane by convection

In the medical field, it is customary to refer to some operating parameters that characterize the transport properties of the whole module. More specifically, the modules are classified according to:

- the value of $K_o A$, where K_o is the overall mass transport coefficient and A the membrane surface area. Obviously, K_o is different for different solutes and depends mainly on the solute molecular size. Usually, the value of $K_o A$ for urea is reported; membranes with urea $K_o A$ values less than 500 ml/min are considered as low-efficiency dialyzers, while high-efficiency dialyzers have urea $K_o A$ values higher that 600 ml/min. It is worth nothing that in order to fully exploit the capacity of high-efficiency modules, both high blood and dialysate fluxes (greater than about 350 and 500 ml/min, respectively) have to be used; if lower flow rates are used, any advantage over low-efficiency modules in terms of performance is lost.
- the ultrafiltration coefficient K_{UF}, that is analogous to the hydraulic permeability; K_{UF} is defined as the ratio of the ultrafiltration rate, Q_F, to the transmembrane pressure difference $(P_B - P_D)$ that drives the ultrafiltration process:

$$K_{UF} = \frac{Q_F}{P_B - P_D} \tag{7.4}$$

where subscripts B and D refer to blood and dialysate, respectively. The US Food and Drug Administration defines low- and high-permeability dialyzers if the K_{UF} values lower of higher than 8 ml h^{-1}mmHg^{-1}, respectively.

In order to quantify the ability of the modules to remove higher MW toxins, the clearance of some reference middle MW molecules (see Table 7.6) is usually indicated by

Table 7.6 Solutes used to test the hemodialysis membrane properties

Solute	MW
Urea	60
Creatinine	113
Uric acid	168
Dextrose	180
Bromosulfophthalein	838
Vitamin B$_{12}$	1355
Insulin	5200
$\beta_2$microglobulin	11800

the producer. The protein β_2-microglobulin, which is involved in the dialysis-related amyloidosis, is often considered as a standard reference to assess the removal of middle MW molecules. High-permeability modules have a β_2-microglobulin clearance (defined in the following section) greater than 20 ml/min. It should be pointed out that the clearance of middle MW molecules depends not only on the features of the module, but also on the operating conditions. In particular, the clearance of middle MW molecules is due to convective transport and depends on the ultrafiltration rate; therefore, a high β_2−microglobulin clearance can be obtained only with membranes with a high ultrafiltration coefficient. Dialyzers with $K_{UF} > 15\,\mathrm{ml\,h^{-1}mmHg^{-1}}$ and β_2−microglobulin clearance exceeding 20 ml/min are considered as high-flux dialyzers, while for low-flux dialyzers K_{UF} and β_2−microglobulin clearance is lower than $15\,\mathrm{ml\,h^{-1}mmHg^{-1}}$ and 10 ml/min, respectively.

7.7 Performance Parameters of a Membrane Module: Clearance and Dialysance

Two main parameters are used to quantify the efficiency of a device used to purify blood in extracorporeal circulation: *clearance* and *dialysance*. Clearance is defined as the ratio of the toxin removal rate from blood to the toxin concentration in the inlet blood (see Fig. 7.8 for symbols):

$$\mathrm{CL} = \frac{Q_{B,in}c_{iB,in} - Q_{B,out}c_{iB,out}}{c_{iB,in}} \tag{7.5}$$

In a similar way, dialysance is defined as

$$\mathrm{DL} = \frac{Q_{B,in}c_{iB,in} - Q_{B,out}c_{iB,out}}{c_{iB,in} - c_{iB,out}^{*}} \tag{7.6}$$

where $c_{iB,out}^{*}$ is the limiting toxin concentration that can be theoretically obtained in the outlet stream, i.e., the toxin concentration at thermodynamic equilibrium with the inlet dialysate stream. The clearance is a parameter that describes the impact of the dialysis process on the patient (in fact, clearance can be considered as the blood volume which should be completely purified in the time unit) and allows to predict the evolution of the toxin concentration in the patient body with time; on the other hand, the dialysance is related to the intrinsic module efficiency and is particularly useful when dialysate regeneration is applied in recirculating devices. Of course, clearance and dialysance coincide if the dialysate fed to the membrane module is free from the solute considered ($c_{iD,in} = 0$), like in the case of single-pass dialysis.

In a pure hemodialysis process, with no significant ultrafiltration rate, $Q_{B,i} = Q_{B,out} = Q_B$; therefore,

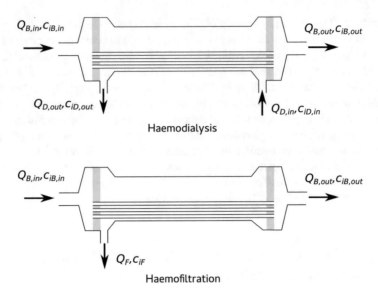

Fig. 7.8 Schema of a membrane module for hemodialysis or hemofiltration

$$CL = Q_B \frac{c_{iB,in} - c_{iB,out}}{c_{iB,in}} \quad DL = Q_{B,in} \frac{c_{iB,in} - c_{iB,out}}{c_{iB,in} - c_{iB,out}^*} \tag{7.7}$$

while in a pure hemofiltration process, with no dialysis fluid fed to the membrane module

$$CL = Q_F \frac{c_{iF}}{c_{iB,in}} \tag{7.8}$$

where Q_F and c_{iF} are the flow rate and the concentration of the ith component in the ultrafiltrate flow, respectively.

In the general case of hemodialysis with significant ultrafiltration rate, inlet and outlet blood flow rates are different ($Q_F = Q_{B,in} - Q_{B,out}$) and clearance and dialysance are given by:

$$CL = Q_{B,in} \frac{c_{iB,in} - c_{iB,out}}{c_{iB,in}} + Q_F \frac{c_{iB,out}}{c_{iB,in}} = CL_0 + Q_F \frac{c_{iB,out}}{c_{iB,in}} \tag{7.9}$$

$$DL = DL_0 + Q_F \frac{c_{iB,out}}{c_{iB,in} - c_{iB,out}^*} \tag{7.10}$$

where CL_0 and DL_0 are defined as the clearance and dialysance that would give the concentration $c_{iB,out}$ in the absence of ultrafiltration.[4]

---

[4]It is worth noting that CL_0 and DL_0 are different from the clearance and dialysance obtained with the same device in the absence of ultrafiltration.

Equation 7.9 can be rearranged as

$$\frac{CL}{Q_{B,in}} = \frac{CL_0}{Q_{B,in}} + \frac{Q_F}{Q_{B,in}}\left(1 - \frac{CL_0}{Q_{B,in}}\right) \tag{7.11}$$

that clearly shows that the clearance improvement due to ultrafiltration is more significant for the components with low CL_0 values.

It is worth to underline that both the dialysance and clearance depend on the module geometry, membrane properties, and operating conditions (mainly blood and dialysate flow rates). Models that describe the solute transport through the membrane and the change in composition of blood and dialysate may be helpful to evaluate the clearance that can be obtained with a dialyzer in different operating conditions; these models can also be used to build a complete pharmacokinetic model to evaluate the effect of a specific treatment protocol on the patient.

7.8 A Simple Model for Hemodialysis

As a first step toward the modeling of a module for renal replacement therapy, let us consider a simple dialysis process, with only diffusive transport of the toxins through the membrane and without ultrafiltration. In this condition, the blood and dialysate flow in countercurrent mode along the module with constant flow rates (see Fig. 7.9). Countercurrent flow ensures the most efficient distribution of the concentration difference (i.e., the driving force for solute permeation) along the module, thus enhancing the performance of the device.

If we assume that, at equilibrium, the concentrations of the solute in blood and dialysate are equal, the flux of solute i, J_i, can be evaluated as follows (see also Chap. 3 for a detailed discussion):

$$J_i = \frac{c_{iB} - c_{iD}}{\frac{1}{k_{B,i}} + \frac{1}{k_{D,i}} + \frac{\delta_m}{\mathscr{P}_i}} = K_{o,i}\,(c_{iB} - c_{iD}) \tag{7.12}$$

Fig. 7.9 Mass balance for a dialysis process without ultrafiltration

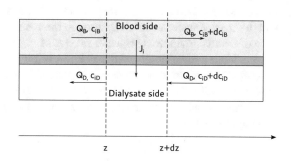

where c_i is the concentration of solute i, k its mass transport coefficient, and $\mathscr{P}$ its permeability in the membrane; subscripts B and D refer to blood and dialysate, respectively. The overall mass transport coefficient can therefore be calculated as follows:

$$\frac{1}{K_{o,i}} = \frac{1}{k_{B,i}} + \frac{1}{k_{D,i}} + \frac{\delta_m}{\mathscr{P}_i} \tag{7.13}$$

The model of the dialyzer is obtained by combining the transfer rate expression (Eq. 7.12) with the solute mass balances in blood and dialysate; for a countercurrent flow, these mass balance equations are written as follows:

$$\frac{dc_{iB}}{dz} = -\frac{K_{o,i}A}{LQ_B}(c_{iB} - c_{iD}) \tag{7.14}$$

$$\frac{dc_{iD}}{dz} = -\frac{K_{o,i}A}{LQ_D}(c_{iB} - c_{iD}) \tag{7.15}$$

where z is the direction along the dialyzer length, A the membrane area, L the module length, and Q_B and Q_D the volumetric flow rate of blood and dialysate, respectively. The above equations can be integrated with the boundary conditions:

$$z = 0 \quad c_{iB} = c_{iB,in}; \quad z = L \quad c_{iD} = c_{iD,in} \tag{7.16}$$

In single-pass hemodialysis (inlet dialysate free of solute), the dimensionless concentrations

$$\tilde{c}_{iB} = \frac{c_{iB,in} - c_{iB}}{c_{iB,in} - c_{iD,out}} \quad \text{and} \quad \tilde{c}_{iD} = \frac{c_{iD} - c_{iD,in}}{c_{iB,in} - c_{iD,out}}$$

can be expressed as a function of two-dimensionless groups:

$$Z = \frac{Q_B}{Q_D} \quad R = \frac{K_{o,i}A}{Q_B} \tag{7.17}$$

where Z depends only on the operating conditions, while R includes also information on the module characteristics. More specifically, R accounts for the membrane area and the mass transfer coefficient through the membrane and may be viewed as the ratio of the maximum toxin transfer rate ($K_o A c_{iB}$) to the toxin axial convection rate ($Q_B c_{iB}$). It is generally assumed that $K_{o,i}A$ is fairly constant over the possible range of dialysate and blood flow rates; however, a deeper analysis shows that the fluid dynamic conditions, and in particular, the thickness of the boundary layer near the membrane surface affects the $K_{o,i}$ value via the mass transport coefficients $k_{B,i}$ and $k_{D,i}$ (see Eq. 7.13).

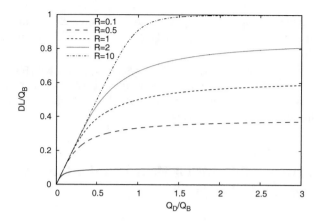

Fig. 7.10 Dialysance as a function of the ratio of dialysate to blood flow rate; *curves* for different values of R are presented

Integration of Eqs. 7.14 and 7.15 provides the dimensionless dialysance as

$$\frac{DL}{Q_B} = \frac{1 - \exp[R(1 - Z)]}{Z - \exp[R(1 - Z)]} \tag{7.18}$$

It is worth noting that the dialysance depends on the module characteristics and the operating conditions, but does not depend on the solute concentration.

Figure 7.10 reports a plot of the dimensionless dialysance, DL/Q_B versus the ratio of dialysate to blood flow rate for different values of R.

The plot clearly shows that:

- Dialysance increases with increasing dialysate flow rate, at constant R and blood flow rate; actually, this is due to the reduction of the toxin concentration in the dialysate and then to the increase in the driving force for the toxin transfer. A limiting dialysance value, DL_∞, is obtained for high dialysate-to-blood flow rates, i.e., for negligible toxin concentration in the dialysate. The limiting value of the dialysance is given by:

$$DL_\infty = Q_B[1 - \exp(-R)] \tag{7.19}$$

From a practical point of view, a negligible improvement of dialysance is obtained by increasing Q_D/Q_B above 1–1.5.
- For low Q_D/Q_B values, DL approaches a value DL_0 such that

$$DL_0 = Q_D \tag{7.20}$$

corresponding to an outlet dialysate stream in equilibrium with the inlet blood. In this limiting case, the dialysance depends only on Q_D/Q_B and not on the value of $K_o A$. In other words, at low dialysate flow rate, the performance of low- and high-efficiency dialyzers (i.e., dialyzers with low or high $K_o A$ values) are similar and the capacity of high-efficiency dialyzers can be exploited only with a high dialysate flow rate.

- Dialysance increases with increasing R values, i.e., by increasing the overall mass transport coefficient K_o and/or the membrane area. Therefore, in order to increase the module dialysance, it is important to reduce the mass transport resistance and use modules with high membrane area.

7.9   Evaluation of the Overall Mass Transfer Coefficient and Module Performance

As previously discussed, the membrane module performance improves with increasing values of the effective membrane area and overall mass transfer coefficient K_o. Equation 7.13 states that K_o depends not only on membrane properties—in particular on the membrane permeability and thickness—but also on the mass transfer coefficients in the blood and dialysate boundary layers; indeed, if a membrane with high permeability is used, mass transport resistances in blood and dialysate may largely affect K_o. As reported in Chap. 2, the mass transport coefficients depend on the fluid dynamic conditions and can be evaluated via dimensionless expressions which relate the Sherwood number ($Sc = k\ell/\mathscr{D}$) to the Reynolds ($Re = \rho v \ell/\mu$) and Schmidt ($Sc = \mu/\rho\mathscr{D}$) numbers. Some correlations suitable for the evaluation of the mass transport coefficients in hollow fiber hemodialyzers are reported in Table 7.7.

It is worth comparing the overall mass transfer coefficients and the clearance of different solutes. To this aim, a case study will be considered here referring to membrane modules with the features reported in Table 7.8; urea, creatinine, and vitamin B_{12} have been chosen to represent small and middle MW uremic toxins. In Table 7.9, membrane permeability and other data for the evaluation of the mass transport coefficients (blood and dialysate side) are reported. It can be seen that the permeabilities of the compounds considered are significantly different (one to two orders of magnitude). As for mass transfer coefficients, the dialysate mass transport coefficient almost doubles the mass transport coefficient in blood; furthermore, the mass transport coefficient for urea is one order of magnitude higher than that of vitamin B_{12}. As a consequence, the membrane permeability is of the same order of magnitude as the mass transport coefficients for urea, while it is one order of magnitude lower than the mass transport coefficient for vitamin B_{12}. Therefore, if we compare the overall mass transport coefficient, K_o, with the membrane permeance, we note that the resistances in blood and dialysate boundary layers significantly affect K_o for urea and creatinine, while transport of vitamin B_{12} is controlled by the membrane permeance. The values of K_o allow to evaluate the clearance of the module: clearance values range from 280 ml/min (corresponding to a detoxification efficiency greater than 90 %) to about 65 ml/min for vitamin B_{12} (corresponding to a detoxification efficiency lower than 25 %). The concentration profiles of the different components along the dialyzer length are reported in Fig. 7.11: the plot shows that

Table 7.7 Correlations for the mass transport coefficients in hemodialysis

Blood side (intralumen flow)		
$Sh = 1.64Gz^{1/3}$	$Gz > 100$	[98]
$Sh = 2.42Gz^{1/4}$	$7 < Gz < 100$	
$Sh = 3.9$	$Gz < 7$	
$Sh = 1.62Gz^{1/3}$	$Gz > 10$	[99]
$Sh = 3.656$	$Gz < 4$	
$Sh = 2.43Gz^{1/3}$	$Gz > 10$	[99]
$Sh = 4.364$	$Gz < 4$	
Dialysate side (extralumen flow)		
$Sh = 1.25\left(Re\,\dfrac{d_e}{L}\right)^{0.93} Sc^{1/3}$	$\phi = 3$ and 23%	[100]
$Sh = 8.8\left(Re\,\dfrac{d_e}{L}\right) Sc^{1/3}$	$\phi = 15\%$	[101]
$Sh = 5.85\,(1-\phi)\,Re^{0.6}Sc^{1/3}\left(\frac{d_e}{L}\right)$	$\phi = 4,\ 8.7,\ 19.7,\ 45\%$	[102]
$Sh = (0.53 - 0.58\phi)\,Re^{0.53}Sc^{1/3}$	$\phi = 31.9$ and 75.8%	[103]

Blood-side correlations are derived from classical Graetz theory with constant wall concentration or linear wall concentration; dialysate-side correlations are derived from Graetz theory, modified with an equivalent diameter ($d_e/4$ is defined as the ratio of the passage area to the wet perimeter) and to account for fiber density (ϕ). Graetz number is defined as $Gz = Re\,Sc\,(d/L)$

Table 7.8 Dialyzer characteristics considered for a case study

Number of fibers	17000
Shell diameter (cm)	4.7
Fiber diameter (cm)	0.02
Membrane thickness (cm)	0.002
Fiber length (cm)	25

urea is removed with a higher rate near the blood inlet section, while creatinine and vitamin B_{12} are more uniformly removed along the dialyzer.

The effect of the dialysate flow rate on the clearance is reported in Fig. 7.12. As previously discussed, an increase in the dialysate flow rate results in an increase of the clearance; nevertheless, for vitamin B_{12}, the increase in the dialysate flow rate has only a minor effect on the dialyzer clearance, which is always less than 70 ml/min.

7.10 A General Model for the Hemodiafiltration Process

In order to obtain a more general model of a dialysis unit, it is necessary to account for the ultrafiltration rate, which determines changes in blood and dialysate flow rate, as well as a convective solute transport that superimposes to diffusion. Indeed,

Table 7.9 Evaluation of the performance of a membrane module (see Table 7.8) with blood and dialysate flow rate of 300 and 500 ml/min, respectively

Membrane

		Urea	Creatinine	B_{12}
$\mathscr{P}/\delta_m$, cm/s		3.26×10^{-3}	0.54×10^{-3}	0.06×10^{-3}

Blood[a]

		Urea	Creatinine	B_{12}
ρ, g/cm^3	1.02			
μ, g cm^{-1}s^{-1}	0.012			
v, cm/s	0.93			
Re	1.58			
$\mathscr{D}$, cm^2/s		7.4×10^{-6}	5.1×10^{-6}	1.2×10^{-6}
Sc		1590	2300	9965
Gz^{c}		2.0	2.9	12.6
k, cm/s^{d}		1.44×10^{-3}	0.99×10^{-3}	0.27×10^{-3}

Dialysate[b]

		Urea	Creatinine	B_{12}
ρ, g/cm^3	1			
μ, g cm^{-1}s^{-1}	0.01			
v, cm/s	0.86			
Re	2.6			
$\mathscr{D}$, cm^2/s		1.85×10^{-5}	1.27×01^{-5}	2.9×10^{-6}
Sc		540	790	3450
Gz^{c}		1.67	2.5	10.8
k, cm/s^{d}		3.61×10^{-3}	2.48×10^{-3}	0.63×10^{-3}

Dialyzer

K_0, cm/s		0.81×10^{-3}	0.31×10^{-3}	0.05×10^{-3}
CL, cm^3/min		280	210	64

[a]Blood properties from [104]
[b]Dialysate properties are set equal to water properties; diffusivities are evaluated by Wilke–Chang method
[c]$G_Z = ReSc \ (d/L)$
[d]Mass transport coefficients are evaluated according to the relationship $Sh = 1.64Gz^{1/3}$ for $Gz \geq 100$, $Sh = 2.42Gz^{0.25}$ for $7 < Gz < 100$, and $Sh = 3.9$ for $Gz \leq 7$

in a membrane with significant hydraulic permeability, the transmembrane pressure difference causes a fluid flux with velocity J_v; then, the solute transport through the membrane is determined by both diffusion and convection. Chapter 3 reports how the solvent and solute fluxes through a membrane are related to hydrostatic pressure, osmotic pressure, and concentration gradients.

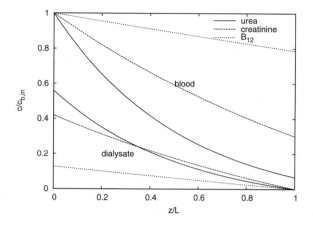

Fig. 7.11 Concentration profiles along the dialyzer length

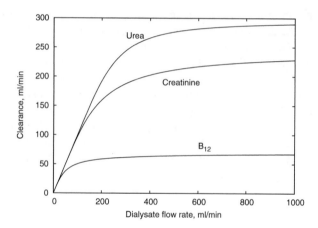

Fig. 7.12 Effect of the dialysate flow rate on the clearance of different solutes

7.10.1 Solvent Transmembrane Flux

The ultrafiltration flux, J_v, is determined by the local transmembrane pressure difference and the membrane hydraulic permeability:

$$J_v = L_P(\Delta P - \Delta \pi) \tag{7.21}$$

where $\Delta \pi$ is the difference in the osmotic pressure between blood and dialysate. Indeed, this difference is due to the oncotic pressure in the blood side, so that the ultrafiltration rate can be written as follows:

$$J_v = L_P[P_B(z) - (P_D(z) + \pi_B(z))] \tag{7.22}$$

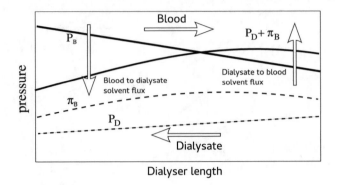

Fig. 7.13 Qualitative plot of the pressure profiles along a hemodialysis module. The figure shows that near the distal section, backfiltration can occur, with reabsorption of dialysate in the blood

It is worth noting that, due to friction losses, the blood and dialysate pressures decrease along the direction of flow, which is opposite for the two streams in countercurrent flow. On the other hand, the blood oncotic pressure, due to the larger molecules rejected from the membrane, increases in the direction of blood flow, at least until water is removed from blood. Therefore, it is possible that the sign of the driving force in Eq. 7.22 changes along the dialyzer length, so that the net fluid flux moves from the blood to dialysate in the proximal section (blood inlet) and form dialysate to blood in the distal section (blood outlet). This phenomenon, called *backfiltration*, obviously affects the solute removal and requires that dialysate be sterile, non-pyrogenic, and endotoxin-free. Figure 7.13 presents a qualitative plot of the hydrostatic and osmotic pressure profiles along the dialyzer length.

7.10.2 Solute Flux

The transmembrane solute flux can be expressed as follows:

$$J_i = J_v \mathscr{S}_i c_i - \mathscr{P}_i \frac{dc_i}{dx} \qquad (7.23)$$

where x is the direction along the membrane thickness. Equation 7.23 highlights the two contributions to solute flux across the membrane: the convective term, proportional to the ultrafiltration rate, and the diffusive term, proportional to the concentration gradient. Integration of Eq. 7.23 across the membrane thickness gives:

$$J_i = \frac{Pe_{tm} \mathscr{P}_i}{\delta_m} \left(\frac{\exp(Pe_{tm})}{\exp(Pe_{tm}) - 1} c'_{iB} - \frac{1}{\exp(Pe_{tm}) - 1} c'_{iD} \right) \qquad (7.24)$$

where δ_m is the membrane thickness and c'_{iB} and c'_{iD} are the solute concentrations at the membrane interface in blood and dialysate, respectively. In Eq. 7.24, the transmembrane Peclet number $Pe_{tm} = \mathcal{S}_i J_v \delta_m / \mathcal{P}_i$ was introduced; such dimensionless number accounts for the relative importance of convective and diffusive transport across the membrane.

It can be easily verified that for a pure dialysis process, with $J_v = 0$ and then $Pe_{tm} = 0$, the solute flux reduces to

$$J_i = \frac{\mathcal{P}_i}{\delta_m} \left(c'_{iB} - c'_{iD} \right) \tag{7.25}$$

whereas for a very high ultrafiltration rate, i.e., for $Pe_{tm} \to \infty$,

$$J_i = \mathcal{S}_i J_v c'_{iB} \tag{7.26}$$

As already pointed out (see Chap. 3), due to the mass transport resistance in the liquid boundary layers, the concentrations at the membrane interfaces are different from those in the bulk of the fluid phase. The relation between the bulk and interface concentrations in blood (upstream phase) is given by

$$c'_{iB} = c_{iB} \left[e^{J_v/k_{B,i}} + \frac{J_i}{J_v c_{iB}} \left(1 - e^{J_v/k_{B,i}} \right) \right] \tag{7.27}$$

If $J_v = 0$ (no ultrafiltration), Eq. 7.27 becomes

$$c'_{iB} = c_{iB} \left(1 - \frac{J_i}{k_B c_{iB}} \right) \tag{7.28}$$

and the solute concentration at the membrane interface is smaller than that in the bulk of the fluid phase (see Fig. 7.14). On the other hand, for a solute completely rejected by the membrane—i.e., for large molecules that have $\mathcal{S}_i = 0$—we have:

$$c'_{iB} = c_{iB} e^{J_v/k_B} \tag{7.29}$$

and the solute concentration at the membrane interface is higher than that in the bulk of the fluid phase. The last phenomenon is known as *concentration polarization*.

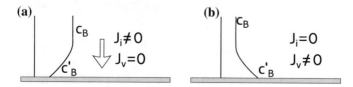

Fig. 7.14 Solute concentration profile in the blood phase: **a** pure dialysis, without ultrafiltration; **b** ultrafiltration with a solute complete rejected by the membrane

As for the dialysate (downstream phase), it is easy to derive that

$$c'_{iD} = c_{iD} \left[e^{-J_v/k_{D,i}} + \frac{J_i}{J_v c_{iD}} \left(1 - e^{-J_v/k_{D,i}} \right) \right] \tag{7.30}$$

By substituting Eqs. 7.27 and 7.30 into Eq. 7.24, a relation for the solute flux in terms of bulk solute concentrations in blood and dialysate is obtained:

$$J_i = \frac{J_v \mathscr{S}_i \left[\exp \left(Pe_{tm} + \frac{J_v}{k_B} + \frac{J_v}{k_D} \right) c_{iB} - c_{iD} \right]}{(1 - \mathscr{S}_i) \exp \left(\frac{J_v}{k_D} \right) \left[\exp \left(Pe_{tm} \right) - 1 \right] + \mathscr{S}_i \exp \left(Pe_{tm} + \frac{J_v}{k_B} + \frac{J_v}{k_D} \right) - \mathscr{S}_i} \tag{7.31}$$

7.10.3   Model of the Membrane Dialyzer

A model to simulate the behavior of a membrane module can be obtained by combining:

- the differential mass balance equations for solvent and toxins (see Fig. 7.15)

$$\frac{dQ_B}{dz} = \alpha \frac{dQ_D}{dz} = -J_v \frac{A}{L} \tag{7.32}$$

$$\frac{d(Q_B c_{iB})}{dz} = \frac{d(Q_D c_{iD})}{dz} = -J_i \frac{A}{L} \tag{7.33}$$

where Q_B and Q_D are for blood and dialysate flow rates and $\alpha = 1$ or -1 for countercurrent or cocurrent flux, respectively;
- the expressions of the solvent (Eq. 7.21) and toxin (Eq. 7.31) flux across the membrane

Fig. 7.15 Scheme for volume and mass balance in a membrane dialyzer

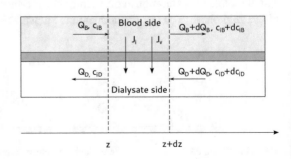

- the momentum balance equation to evaluate the local pressure, in the blood and dialysate compartment. In hollow fiber dialyzers, the pressure drop can be evaluated with the Hagen–Poiseuille equation (for intralumen flow) or its modification (for flux in the shell)

The above-described set of differential equations is completed by trivial boundary conditions, which account for flow rate, composition, and pressure of the incoming blood and dialysate streams; such boundary conditions are not reported for the sake of brevity. In general, the so posed problem requires a numerical solution, even if some analytical solutions are obtained for limiting cases or by using simplifying assumptions. Here, in order to clarify the behavior of a renal replacement device operating in hemodiafiltration conditions, we report the results of some simulations [105] referred to a membrane module with the characteristics summarized in Table 7.10. The operating conditions considered for the simulations are reported in Table 7.11. Figure 7.16 shows the concentration profiles along the device for three compounds, chosen as representative of small and middle MW molecules (the clearance of urea is very similar to that of creatinine and is not reported for the sake of legibility). The plot clearly shows that the module is able to significantly reduce the concentration of middle MW compounds, even if removal of creatinine is more efficient. The effect

Table 7.10 Characteristics of the high-flux dialyzer considered for simulation

Fiber number	16666	Hydraulic permeability	cm/Pa s	1.4×10^{-8}
Fiber density, ϕ	0.55	Solute	Membrane permeability	Sieving
Fiber diameter (μm)	190		$\mathscr{P}/\delta$, cm/s	
Membrane area (m^2)	2.48	Urea	9.5×10^{-4}	1
Fiber length (cm)	25	Creatinine	6.5×10^{-4}	1
Shell diameter (cm)	4.65	Vitamin B$_{12}$	2.5×10^{-4}	1
		β_2-microglobulin	1.5×10^{-4}	0.63

Table 7.11 Operating conditions used for the simulations of a hemodiafiltration process carried out with a high-flux dialyzer

	Blood	Dialysate
Inlet flow rate, ml/min	300	500
Outlet pressure, Pa (gauge)	600	0
Concentration, mg/ml		
Urea	2.3	0
Creatinine	0.14	0
Vitamin B$_{12}$	0.035	0
β_2-microglobulin	0.035	0
Proteins	70	0

(Module properties are reported in Table 7.10)

Fig. 7.16 Concentration
profile of creatinine (Cr),
vitamin B$_{12}$ (B12), and
β_2-microglobulin (β_2M)
along the membrane module
with the characteristics
reported in Table 7.10. The
operating conditions
considered are reported in
Table 7.11

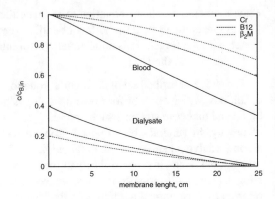

of ultrafiltration is shown in Fig. 7.17, where the flow rates of blood and dialysate along the length of the module are reported. The blood flow rate decreases along the direction of flow in the first part of the device, near the proximal section, while it increases near the distal end. Indeed, in the first part of the membrane module, there is a net fluid flux from blood to dialysate, while in the second part backfiltration occurs and the water flux is reversed. Such a behavior was already discussed in Sect. 7.10.1 on a qualitative basis; here, the quantitative pressure profiles are reported for the specific case under consideration (Fig. 7.18). The driving force for the volume flux is given by the difference $P_B - (P_D + \pi_B)$: as shown in Fig. 7.18, in the first part of the module P_B is higher than $P_D + \pi_B$ and the net water flux is directed from blood to dialysate; moving along the blood flow direction, the driving force decreases (in this simulation, mainly as a result of the pressure drop in the blood channel) and becomes zero at about 13 cm from the inlet section. Starting from this point, the difference $P_B - (P_D + \pi_B)$ is negative and drives the flux from dialysate to blood.

Figure 7.19 reports the effect of the blood and dialysate flow rates on the clearance of the four reference compounds considered. It is interesting to note that the clearance increases almost linearly with the blood flow rate when the dialysate flow rate is kept constant; however, it is worth to note once again that the blood flow rate is limited by the capacity of the vascular access. The clearance also increases if the dialysate flow rate is increased by keeping the blood flow rate constant, but in this case a limiting value is attained.

7.11   Patient Device Models

In the previous section, we discussed the performance of a membrane module in relation to its characteristics and operating conditions.

Actually, the performance of the membrane module is only one of the parameters that interplay to affect the outcome of a renal replacement therapy. Indeed, such a therapy is aimed at replacing the function of the failing kidneys by removing

Fig. 7.17 Profile of blood and dialysate flow rates and ultrafiltration rate along the length of the membrane module. The results refer to a module with the characteristics reported in Table 7.10 under the operating conditions reported in Table 7.11

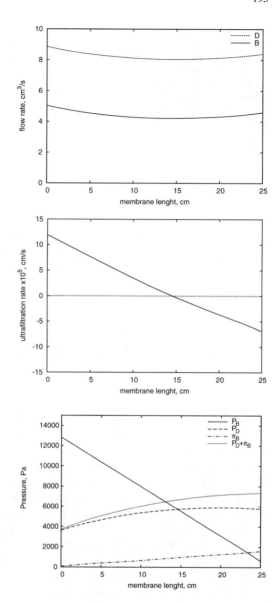

Fig. 7.18 Pressure profiles along the length of the membrane module with the characteristics reported in Table 7.10. The results refer to a module with the characteristics reported in Table 7.10 under the operating conditions reported in Table 7.11

metabolic waste products and regaining control over the level of toxic substances in plasma (mainly the small uremic toxins but also middle and high MW compounds). Obviously, the fundamental difference between the detoxification action of healthy kidneys and a chronic hemodialysis therapy is in the intermittent nature of the latter, which is characterized by a fast intradialytic decrease in toxin concentrations followed by toxin accumulation in the interdialytic period. Furthermore, in dialysis, toxins are removed only from the hematic compartment (the one that is actually

Fig. 7.19 Effect of blood (*above*) and dialysate (*below*) flow rate on the clearance of four reference solutes. The results refer to a membrane module with the characteristics reported in Table 7.10. The inlet stream composition is reported in Table 7.11.

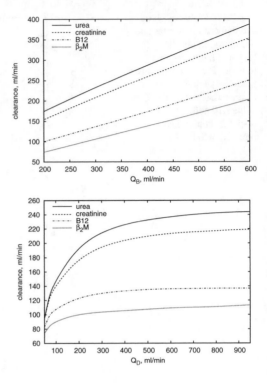

perfused through the membrane module) and the overall detoxification efficiency may be reduced if toxin transfer from other compartments such as tissues is slow compared to the duration of the treatment. Therefore, duration and frequency of the dialysis sessions strongly affect the time course of the plasma level of toxins and, thus, the effectiveness of the treatment. Indeed, it is generally accepted that the evolution of the hematic levels of specific toxins during and after a renal replacement therapy is the basis of its adequacy and has a strong impact on the quality of life of the patient and on the final outcome of the treatment.

In this context, the utility of a patient-device model (following the same approach of *pharmacokinetic models*) able to predict the intradialytic and interdialytic toxin level evolution is apparent. Such models can be useful to design not only the devices, but also the therapy protocol itself and help in taking clinical decisions about duration and frequency of the dialysis sessions and, in general, about the management of a patient in dialysis.

Several pharmacokinetic models applied to renal replacement therapies are reported in the literature (see, for example, [106–108]). All such models are based on a *compartmental* approach, i.e., the human body is described as a set of different well-mixed compartments, which can exchange solutes by passive transport. In each compartment, mass balances of some target solutes provide the concentration evolution both during the treatment session and in the interdialytic period; the rates of solute generation/consumption by metabolic processes and exchange with other compartments are accounted for in the mass balances.

In the following sections, two examples of the application of pharmacokinetic models to renal replacement therapy are discussed. Firstly, a single-compartment model, which can be considered as the simplest option, is presented: the application of this model leads to the identification of a dialysis dosage parameter that provides some information about the adequacy of the treatment. Subsequently, a more complex model is presented, which accounts for a non-perfused and a perfused compartment in the patient body and provides a better description of the toxin concentration evolution.

7.11.1 Single-Compartment Model

In a single-compartment model, the patient is described as a single, homogeneous compartment of volume V (see Fig. 7.20). The solute is generated in the compartment and removed by an endogenous clearance, due to the residual renal function and/or to a non-renal process; furthermore, during the dialysis sessions, the dialyzer clearance is also accounted for.

The solute balance over the compartment yields the following differential equation (the subscript i for properties related to a single component is omitted for the sake of brevity)

$$\frac{d}{dt}(Vc) = G - K_r c - CL(c - c_D) \tag{7.34}$$

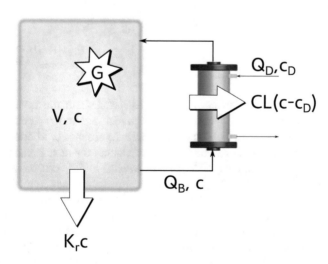

Fig. 7.20 Conceptual scheme of a single-compartment model

where c is the solute concentration in the pool (equal to the concentration in blood), G the generation rate, K_r the endogenous (or residual clearance), CL the dialyzer clearance, and c_D the solute concentration in the dialysis fluid. For toxins such as urea and creatinine, usually $c_D = 0$; if a constant distribution volume is considered, the mass balance equation can be simplified in:

$$V\frac{dc}{dt} = G - Kc \tag{7.35}$$

with $K = \text{CL} + K_r$.

Integration of Eq. (7.35) with the initial condition $c(t = 0) = c_0$ gives

$$c = c_0 e^{-Kt/V} + \frac{G}{K}\left(1 - e^{-Kt/V}\right) \tag{7.36}$$

Equation 7.36 provides the time course of concentration:

- during the dialysis phase, for $t \in [0, t_{dia}]$ (t_{dia} is the duration of the dialysis session), with the initial concentration equal to the predialytic concentration ($c_0 = c_{pre}$) and $K = K_{dia} = K_r + \text{CL}$
- during the interdialytic period, for $t \in [t_{dia}, t_{dia} + t_{id}]$, with

$$c_0 = c_{pre} e^{-K_{dia} t_{dia}/V} + \frac{G}{K_{dia}}\left(1 - e^{-K_{dia} t_{dia}/V}\right) \text{ and } K = K_r.$$

In a periodic renal replacement therapy, the concentration at the end of the interdialytic period is the concentration at the beginning of the dialysis; this condition allows to derive the following expression for c_{pre}:

$$c_{pre} = \frac{G\left[K_r e^{-K_r t_{id}/V}\left(1 - e^{-K_{dia} t_{dia}/V}\right) + K_{dia}\left(1 - e^{-K_r t_{id}/V}\right)\right]}{K_{dia} K_r \left(1 - e^{-(K_{dia} t_{dia} + K_r t_{id})/V}\right)} \tag{7.37}$$

Equation 7.37 can be used to analyze clinical data and obtain the model parameters related to the patient (V, G, K_r); once these parameters are known, the model can be used to predict the effect of a dialytic treatment. In the absence of a better estimation, standard values for V and G can be used: in particular, for urea V can be assumed equal to the total body water (58 % of body weight) and G can be related to the protein intake: $G = 0.11\,I - 0.12$, where I is the protein intake in mg/min.

Plots of the concentration time course predicted with a single-compartment model in selected sets of conditions (see Table 7.12) are reported in Figs. 7.21 and 7.22. All the simulation runs refer to a dialysis protocol with three sessions per week, and the following parameter values were assumed: $V = 40\,\text{l}$, $G = 6.25\,\text{mg/min}$.

Figure 7.21 shows that an increase in either the dialysis clearance or the duration of the session results in a decrease of urea concentration; indeed, the concentration profile is mainly controlled by the value of the dimensionless group $K t_{dia}/V$, so that by varying K or t_{dia} by the same factor a similar result is obtained (see curve B and

Table 7.12 Operating conditions considered for the concentration profiles reported in Figs. 7.21 and 7.22

Condition set	t_{dia}, h	CL, ml/min	K_r, ml/min	Kt_{dia}/V	TAC, mg/dl	EKR, ml/min
A	3	200	0	0.9	61.8	10.1
B	3	266	0	1.2	48.4	12.9
C	4	200	0	1.2	48.2	13.0
D	4	250	0	1.5	40.4	15.4
Ca	4	200	5	1.23	35.6	17.6
Cb	4	200	10	1.26	28.1	22.2

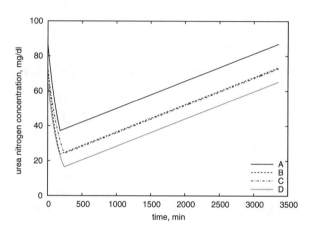

Fig. 7.21 Concentration time course predicted with a single-compartment model. The different *curves* correspond to different sets of conditions according to Table 7.12

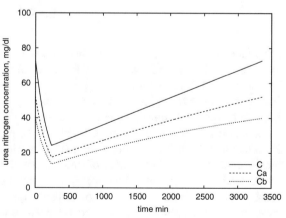

Fig. 7.22 Concentration time course predicted with a single-compartment model for different residual renal clearances (operating conditions according to Table 7.12)

C in Fig. 7.21). The effect of the group Kt_{dia}/V on the time-averaged concentration (TAC), defined as

$$\text{TAC} = \frac{1}{t_{dia} + t_{id}} \int_0^{t_{dia}+t_{id}} c(t)\, dt \qquad (7.38)$$

Fig. 7.23 Effect of the dose per session on the average urea nitrogen concentration (TAC) and the equivalent renal clearance (EKR). *Curves* refer to $V = 40$ l, $G = 6.25$ mg/min, $K_r = 0$; dialysis protocol: three-weekly sessions of 4 h each

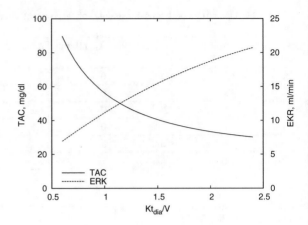

and on the equivalent renal clearance (EKR), defined as

$$EKR = \frac{G}{TAC} \tag{7.39}$$

is reported in Fig. 7.23.

According to the single-compartment model, the parameter Kt_{dia}/V, often referred to as *dialysis dose*, has been widely used to quantify the adequacy of dialysis processes. Indeed, if we neglect the solute generation during the dialysis session, this parameter can be easily calculated from Eq. 7.35 as

$$\frac{Kt_{dia}}{V} = -\ln\frac{c_{post}}{c_{pre}} \tag{7.40}$$

If the dialysis time is fixed, the idea is that the clearance and the duration of the dialysis must be adjusted to achieve a target Kt_{dia}/V value. For three-weekly sessions, dialysis doses for urea greater than 1.2 per sessions are widely accepted as adequate and short dialysis times could be permitted as long as this target is attained. Even if this approach to therapeutic planning is attractive for its simplicity, its validity is questionable at least for two reasons. Firstly, referring only to urea removal may be too restrictive, even if this compound can be considered as a good representative for all small solutes: the behavior of middle MW solutes and small proteins may not be well represented and the effect of changing t_{dia} or K, while maintaining a constant dialysis dose, is different for different types of molecules. Furthermore, the assumption that the human body acts as a single well-mixed space is clearly an oversimplification, especially when solutes are promptly removed from blood.

7.11.2 Two-Compartment Model

A better description of the solute removal in a renal replacement therapy requires to consider that in the human body toxins are distributed in different compartments (intracellular fluids, extracellular fluids, blood, etc.), which exchange fluid and solutes with a finite rate, while dialysis removes the solutes only from the compartment that actually perfuses the dialyzer (i.e., blood). Even if the solute removal from this compartment is fast, sequestration of solutes occurs in the other compartments and the overall solute removal is reduced; actually, the solute concentration in plasma falls promptly during the treatment sessions, while the concentration in the other compartments may remain unaffected, due to the low intercompartmental mass transfer rate. This behavior may result in the so-called *postdialysis rebound* in plasma solute concentration, i.e., the release of solutes from the non-perfused compartments to blood, which causes a sharp increase of the blood levels of toxins in the early postdialysis period. A two-compartment model may therefore be more suitable to obtain a more accurate description for the solute removal process and to correctly predict specific phenomena such as the postdialysis rebound.

In the two-compartment model, the total distribution volume is divided into an intracellular or non-perfused compartment, of volume V_{NP}, and an extracellular or perfused compartment (which is directly affected by the dialysis treatment) having a volume V_P. Figure 7.24 shows a conceptual scheme of the model considered. The definition of the two compartments has to be considered strictly in relation to the model development and not necessarily to a physiological interpretation; furthermore, the definition of these compartments and of their volume may be different for different solutes.

Mass exchange between the non-perfused and perfused compartments occurs according to the concentration gradient of individual solutes, with an intercompartmental mass transport coefficients $K_{c,i}$. It is also assumed that solute generation and residual endogenous clearance occur only in the perfused compartment. During the

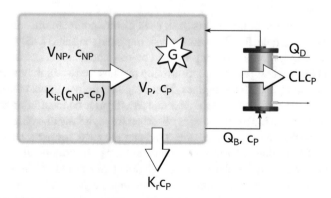

Fig. 7.24 Conceptual scheme of a two-compartment model

dialysis sessions, a bloodstream of flow rate Q_B is circulated through the membrane module, and for a given solute i, a clearance CL_i is obtained, which depends on the ratio Q_B/Q_D and $K_{o,i} A$, as previously discussed.

If the distribution volumes are assumed as constant and the inlet dialysis fluid is free of toxins, for a pure dialysis process the mass balance equations for the two compartments can be written as follows:

• non-perfused compartment

$$\frac{dc_{NP,i}}{dt} = -\frac{K_{c,i}}{V_{NP}} \left(c_{NP,i} - c_{P,i}\right) \tag{7.41}$$

• perfused compartment

$$\frac{dc_{P,i}}{dt} = \frac{K_{c,i}}{V_P} \left(c_{NP,i} - c_{P,i}\right) + \frac{G_i}{V_P} - \frac{(K_r + CL_i)}{V_P} c_{P,i} \tag{7.42}$$

where CL_i is evaluated according to the dialyzer model in the intradialytic period and is set to zero in the interdialytic period.

Equations 7.41 and 7.42 can be integrated with an initial condition on the concentrations $c_{NP,i}(t = 0)$ and $c_{P,i}(t = 0)$; in a periodic dialysis regime, such initial conditions may be replaced by setting the concentration at the end of an interdialytic period equal to the predialytic concentration.

Some examples of compartment concentration time courses in intradialytic and early postdialytic periods are reported in Fig. 7.25: the salient features enlightened in the plots are the presence of an intercompartmental concentration difference during the intradialytic period and postdialytic rebound in the perfused compartment in the early postdialytic time. Both effects are quite small for urea, but more evident for creatinine and even more for β_2-microglobulin. This behavior is related to the different ratios of the intercompartmental exchange coefficient to the dialyzer clearance: this ratio is about 2.9 for urea, but drops to 1.6 for creatinine and 1.4 for β_2-microglobulin.

The two-compartment model has been used also to compare different treatment protocols: as an example, five different dialysis regimes (see Table 7.13) are examined and the concentration time courses predicted by the model are reported in Fig. 7.26. The plots on the left compare the concentration profiles for two high flow rate (A, B) and slow, low-flux (C) dialysis regimes. These regimes deliver almost the same dose (Kt_{dia}/V) per week of treatment for urea; indeed, the urea concentration profiles are very similar, even if a higher rebound is observed for high flow dialysis; in any case, all the regimes achieve almost the same TAC and the EKR for urea. Differences in the concentration profiles are much more evident for β_2−microglobulin; in this case, large postdialytic rebound is observed for high flow regimes and low flow regime (C) achieves better performance in terms of time-averaged concentration and equivalent renal clearance. Plots on the right compare regimes that differ in the frequency of the treatment, though the total weekly session duration is almost the same; in particular,

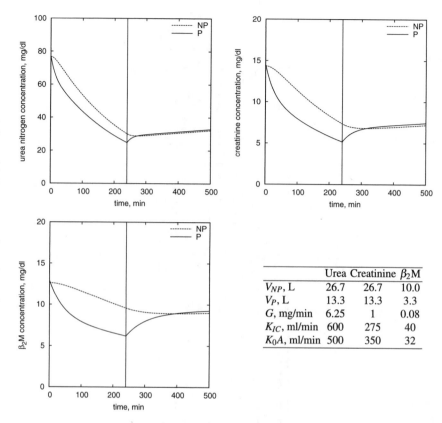

Fig. 7.25 Concentration time courses of urea, creatinine and β_2−microglobulin, predicted by a two-compartment model. Simulations carried out according to dialysis conditions A in Table 7.13. For the sake of clarity, only the intradialytic and early postdialytic periods are reported. Values of model parameters are taken from [109, 110].

	Urea	Creatinine	β_2M
V_{NP}, L	26.7	26.7	10.0
V_P, L	13.3	13.3	3.3
G, mg/min	6.25	1	0.08
K_{IC}, ml/min	600	275	40
K_0A, ml/min	500	350	32

regime E corresponds to a short daily treatment, as in the case of home dialysis. The three regimes A, B, and E achieve almost the same performance in terms of TAC and EKR (regime E gives worse TAC and EKR values, but has a slightly lower weakly treatment time and, therefore, dose per week), but more frequent short dialysis sessions result in more uniform toxin concentration time course in the interdialytic period.

7.12 Regenerative Dialysis

Since the 1960s, hemodialysis is an established treatment for patients with end-stage renal disease; however, in spite of large technological improvements in the last fifty years, the therapy still suffers from several drawbacks. The typical hemodialysis

Table 7.13 Performance of different dialysis regimes

	Frequency (per week)	Duration (h)	Q_B (ml/min)	Q_D (ml/min)	Clearance (ml/min)			Kt_{dia}/V (per week)	TAC (mg/dl)			EKR (mil/min)		
					Urea	Creat	β_2M	Urea	Urea	Creat	β_2M	Urea	Creat	β_2M
A	3	4	350	350	206	175	29	3.7	52.4	10.5	10.7	11.9	9.5	1.6
B	3	3.5	450	450	237	197	30	3.7	53.1	10.9	12.0	11.7	9.2	1.4
C	3	7	150	150	115	105	26	3.6	50.2	9.2	6.8	12.4	10.8	2.5
D	4	3	350	350	206	175	29	3.7	50.8	10.2	10.4	12.3	9.8	1.6
E	7	1.5	350	350	206	175	29	3.2	55.3	10.9	11.0	11.3	9.1	1.5

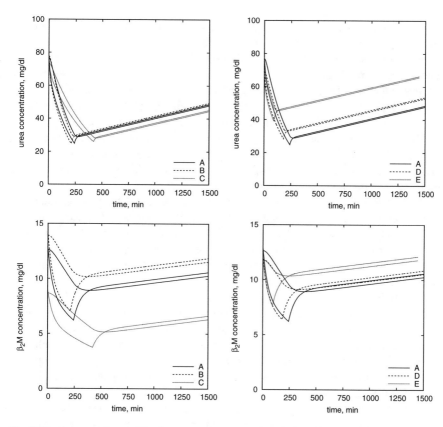

Fig. 7.26 Concentration profiles for urea and β_2-microglobulin in different dialysis regimes (see Table 7.13). For each regime, the *upper curve* in the intradialytic interval refers to the non-perfused compartment, whereas the *lower curve* refers to the perfused compartment. For the sake of clarity, only the intradialytic and the early postdialytic (until 1500 min) periods are reported; in the subsequent interdialytic period, the concentrations rise linearly up to the predialytic concentration

protocols include three-weekly dialysis sessions, carried out in a hospital or specialized dialysis units; even with a short session duration of 4 h, the therapy has disadvantages from a psychosocial point of view, because the patients are tethered to an essentially unmovable machine for a long time. On the other hand, there is much experimental evidence that increased dialysis frequency and prolonged sessions result in biochemical and clinical improvements, with reduced complications and a better outcome in terms of quality of life. In particular, with longer sessions, solutes are more efficiently removed from all the compartments of the patient's body. Prolonged dialysis sessions appear to be necessary to effectively remove the middle MW molecules, which are characterized by slow diffusion and poor clearance in the membrane module and are primarily present in the intracellular volume. A similar problem is phosphate retention, which is frequent in hemodialysis patients: even if membrane modules show a high phosphate clearance, phosphate accumulates mainly in the intracellular space and intercompartmental transport limits the rate of removal of this component; there is clinical evidence that daily prolonged dialysis could help to control serum phosphate concentration. Furthermore, longer sessions with slow rate of fluid removal and improved volume control result in improved blood pressure control and reduce the cardiac complications. Indeed, the natural kidneys work 24 h a day, 7 days a week, and it is quite obvious that daily dialysis would be closer to the actual physiological conditions, with strikingly improved outcomes.

However, more frequent and longer dialysis sessions meet with serious problems, including the need for more dialysis units, with appropriate manpower in nurses and technicians; moreover, patient compliance issues, due to the negative impact on the patients lifestyle, are also a major drawback.

Therefore, there is a need for alternative dialysis systems, to allow patients with chronic kidney failure to be independent from a dialysis center and/or access to more frequent dialysis. A versatile, portable dialysis device could at least partially address these issues; such a device could be used in a variety of environments, as the patient's home or on vacation, as well as anywhere in the hospital from an intensive care unit or in the room of an isolation ward.

One of the problems to solve in order to design a transportable hemodialysis device is related to the need of the large volume of "dialysis-grade" water (at least 120 l per treatment) for the preparation of the dialysate; this requires a continuous water source capable of delivering at least 1 l/min, a water purification system, including a multilayered prefiltration and reverse osmosis unit, sufficient to provide 500–800 ml/min of purified water, a drain system to collect the spent dialysate and the filtrated water from reverse osmosis.

As an alternative, *sorbent dialysis*[5] has been developed as a system that continuously regenerate and reuse a few liters of dialysate. A scheme of a sorbent dialysis system is reported in Fig. 7.27. The core of the regenerative dialysis process is the sorbent unit that must be able to remove a variety of chemical substances from the spent dialysate, maintain the appropriate concentration of electrolytes, and control acidosis. Activated carbon is the first choice sorbent material, since it is able to remove organic compounds, middle MW molecules, uric acids, creatinine, and heavy metals; basically, all organic uremic toxins are removed by activated carbon, with the only notable exception of urea. Actually, activated carbon adsorption capacity for urea is very small and no suitable sorbent is available for this toxin [111]. Apparently, the best solution is to convert urea to ammonium and carbonate using urease as catalyst and remove ammonium ions by a cation exchange material (zirconium phosphate).

The currently used sorbent dialysis systems use a four-layer sorbent cartridge [112] containing:

1. a purification layer containing activated carbon, to adsorb the organic metabolites from the patient and other compounds from the feed water.
2. an enzyme layer consisting of immobilized urease to convert urea to ammonium carbonate according to the reaction

---

[5]The term "sorbent dialysis" is widely used in the literature; as it is reported below, it must be pointed out that regeneration is not exclusively carried out by sorption processes.

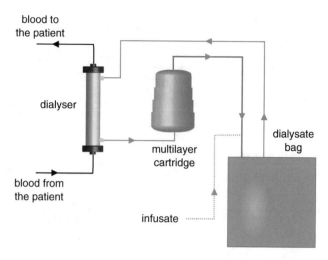

Fig. 7.27 Scheme of a sorbent dialysis system. Pumps, sensors, and control system are not reported

$$(NH_2)_2CO + H_2O \rightarrow (NH_4)_2CO_3$$

A layer of alumina is also included.

3. a cation exchange layer consisting of zirconium phosphate in the Na^+ and H^+ form (NaHZrP) to bind ammonium and to convert carbonate to bicarbonate; this layer also removes Ca^{2+}, Mg^{2+}, K^+, and other cations that may be present in the dialysate. The ratio of Na^+ to H^+ ions can be controlled by preliminary treating of ZrP with a solution of NaOH.
4. an anion exchange layer of zirconium oxide to remove phosphate from uremic patients and other anions present in tap water; in older cartridges, ZrO was used in acetate form; in newer models, sodium zirconium carbonate is also present as a way to supplement acetate and bicarbonate ions while removing undesired anions.

Some typical characteristics of dialysate regeneration cartridges are reported in Table 7.14.

At the end of the regeneration process, the outlet fluid is a water solution containing Na^+, H^+, HCO_3^-, and acetate ions and is mixed with a fresh concentrate containing K^+, Ca^{++}, and Mg^{++} before being recirculated to the dialyzer. Usually, 6 l of dialysis solution is used; since the sorbent cartridge acts also as a filter for bacterial contamination and endotoxin adsorption, tap water can be used. It is worth noting that the use of small dialysate volumes can present some clinical advantages; while in single-pass dialysis the dialysate volume is three times the patient body fluid so that the patient body chemistry is forced to match the dialysate prescription, in regenerative dialysis, the dialysate volume is much smaller than the patient body fluid and the patient is able to control the dialysate.

Table 7.14 Characteristics of sorbent cartridges used in sorbent dialysis systems [113]

	Remove	Release	Sorbent mass, g		Capacity
			4-h treatment	8-h treatment	
Activated carbon	Heavy metals, oxidants, chloramine, creatinine, uric acid, middle molecule, etc.		250	250	
Urease/alumina	Urea	Ammonium carbonate	250	200	20–30 g urea hydrolyzed
Alumina			305	305	
ZrP	NH_4^+, Ca^{2+}, Mg^{2+}, K^+	Na^+, H^+	1300–1600 (pH 5.5)	1400–1700 (pH 5.75–6)	~30 mg NH_4/g ZrP
HZO.Ac/NaZC	Phosphate, fluoride, heavy metals	Acetate, bicarbonate	200 (1:1)	200 (1:1)	30–35 mg PO_4/g NaZC

Data refer to use in hemodialysis

The first sorbent dialysis system, called REDY (Organon Teknika, Oss, The Netherlands)—an acronym for REcirculation of DialYsate—was placed on the market in 1973 and was progressively improved with chemical refinements and system design [114]; while the REDY system was widely used worldwide—during the 1988 earthquake in Armenia, international dialysis teams participating in the medical relief mission made extensive use of it—its production was discontinued in 1993. In recent years, interest in sorbent-based systems has rekindled and the Alliant® system (Renal Solutions, Warrendale, PA), reengineered from the original REDY system, was developed at the beginning of the 2000s [115]. Notably, in this system blood flow is provided by a pressure-actuated diaphragm pump—pulsar blood movement system—which applies the desired pressure gradient to the blood entering and leaving the system, without exposing blood to high shear rates.

7.12.1 Model of a Regenerative Dialysis Process

In this section, a model is presented and used to discuss the effectiveness of a regenerative dialysis system and compare its performance with single-pass dialysis. In terms of blood detoxification capacity, it is obvious that a regenerative system is inferior to single-pass dialysis, because the dialysate regeneration is never complete and some

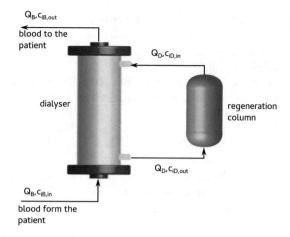

Fig. 7.28 Scheme of a regenerative dialysis system

toxin content is always present in the dialysate fed to the membrane module. On the other hand, the apparent advantages for the patients deriving from increased mobility, and freedom from large water sources and treatment systems increase the interest for the regenerative dialysis; indeed, these advantages also allow to provide the patients with prolonged dialysis. Development of this technology is founded in the research for new, more efficient sorbent materials and a proper design of the whole device. Proper design of these systems is aimed at sizing the regeneration units and choosing the minimal amounts of sorbents to ensure correct operation without breakthrough of toxins throughout the session.

As a first approach, let us consider the clearance that can be obtained using a dialysis membrane module with continuous regeneration of the dialysate (see Fig. 7.28). According to the symbols reported in the figure, we have

$$\mathrm{CL}_i = Q_B \frac{c_{iB,in} - c_{iB,out}}{c_{iB,in}} \qquad \mathrm{DL}_i = Q_B \frac{c_{iB,in} - c_{iB,out}}{c_{iB,in} - c_{iD,in}} \qquad (7.43)$$

where the dialysance depends on the ratio (Z) of blood to dialysate flow rates and on the capacity of the membrane module ($R = K_o A / Q_B$) according to Eq. 7.18.

In a quasi-steady-state mode, it can be easily proved that (see Chap. 8)

$$\frac{Q_B}{\mathrm{CL}_i} = \frac{Q_B}{\mathrm{DL}_i} + \frac{Q_B}{Q_D} \frac{1 - \eta_{i,reg}}{\eta_{i,reg}} \qquad (7.44)$$

where the regeneration effectiveness for component i, $\eta_{i,reg}$ is defined as

$$\eta_{i,reg} = \frac{c_{iD,out} - c_{iD,in}}{c_{iD,out}} \qquad (7.45)$$

The clearance of the regenerative system depends both on the dialysance of the membrane module and on the efficiency of the regeneration system. Indeed, for a high regeneration performance ($\eta_{i,reg} \rightarrow 1$) $CL_i \simeq DL_i$ while for a scarce regeneration efficiency ($\eta_{i,reg} \rightarrow 0$) $CL_i \simeq Q_D \eta_{i,reg}$.

As for the urea, regeneration occurs only in the enzyme layer, where the hydrolysis reaction is carried out; if first-order kinetics and plug flow in the column are assumed

$$\eta_{urea,reg} = 1 - \exp\left(-\frac{k_E M_E}{Q_D}\right) = 1 - \exp\left(-\frac{k_E M_E}{Q_B} Z\right) \qquad (7.46)$$

where k_E is the kinetic constant per unit amount of enzyme (urease) and M_E the enzyme amount in the reactor.

Therefore, we obtain

$$\frac{Q_B}{CL_{urea}} = \frac{Q_B}{DL_{urea}(Z, R)} + Z \frac{\exp\left(-\frac{k_E M_E}{Q_B} Z\right)}{1 - \exp\left(-\frac{k_E M_E}{Q_B} Z\right)} \qquad (7.47)$$

The above equation shows that the clearance of the regenerative dialysis system depends on R, Z, and the parameter

$$\alpha_E = k_E M_E / Q_B \qquad (7.48)$$

A higher enzyme activity (i.e., higher $k_E M_E$) allows a better dialysate regeneration, so that the clearance of the regenerative system approaches that of single-pass dialysis. The dimensionless clearance of the regenerative system is plotted against the ratio Q_D/Q_B in Fig. 7.29. It is worth noting that, for a fixed enzyme load in the column, the regeneration effectiveness for urea decreases as the dialysate flow rate increases; in the limit for $Q_D/Q_B \rightarrow \infty$, the urea clearance tends to the value

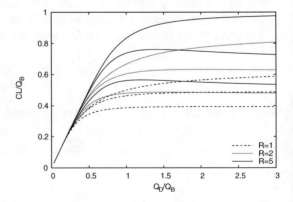

Fig. 7.29 Urea clearance in a regenerative dialysis unit. *Curves* for the same R value refer to different α_E values (the *upper curve* corresponds to the clearance in the single-pass mode (or $\alpha_E \rightarrow \infty$), while the lower curves correspond to $\alpha_E = 2$ and $\alpha_E = 1$)

$$\frac{CL_{urea}}{Q_B} = \frac{\alpha_E \left(1 - e^{-R}\right)}{\alpha_E + 1 - e^{-R}} \tag{7.49}$$

For complete regeneration ($\alpha_E \to \infty$), Eq. 7.49 converges to the expression of the limiting clearance of single-pass dialysis (Eq. 7.19) for high dialysate flow rates.

As for the removal of the ammonium ions produced from the hydrolysis of urea, a cation exchange unit is included (ZrP). For such a unit, if the characteristic time of the solute transfer from the liquid to the solid phase is shorter than the residence time in the column, a sharp breakthrough curve is obtained and an ammonium-free effluent is obtained up to the breakthrough point. Actually, experimental evidence shows that a saturation front is obtained with the available cartridges. From an engineering point of view, this means that the true problem is to design a cation exchange layer large enough to work through the entire treatment session, avoiding the layer saturation.

7.13 Wearable Artificial Kidney

The ultimate goal in the development of any supportive or substitutive device to be applied in chronic organ failure is to obtain an implantable system. Unfortunately, as far as artificial kidneys are concerned, such a significant achievement for the quality of life of patients is not foreseeable in the near future. However, a medium-term objective may be the production of an external wearable device able to clean blood continuously like the natural kidneys, small and light enough to allow patients to ambulate and perform daily-life activities.

It is interesting to note that the first attempt to realize a wearable artificial kidney dates back to the 1970s, when Kolff designed a wearable device and used it in a program which he called "Dialysis in Wonderland" [116]; during such program, several patients were sent in trips in some US locations, where they could enjoy everyday life activities and receive a daily dialysis dose, even without having to go in dialysis centers.

Noways, research on the wearable artificial kidney (WAK) is mainly devoted to the miniaturization of a sorbent dialysis system and has to face some major challenges:

- a true WAK has to be independent of a stationary electrical source and must have a small and light battery to provide all the energy required for its operation.
- the dialysate volume must be further reduced to avoid the device to be excessively weighty.
- small and highly efficient membrane modules and sorbent cartridges are required.
- small and light pumps are required to circulate liquids streams, dose-different components of the dialysate, and control ultrafiltration.

Impressive results have been obtained by Gura et al. [117, 118] with a wearable device intended to be worn continuously as a belt (WAK). The device (see Fig. 7.30 for a scheme and Fig. 7.31 for a picture) includes a standard commercial membrane

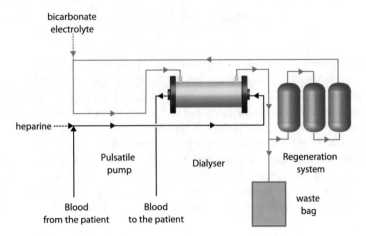

Fig. 7.30 Scheme of the wearable artificial kidney developed by Gura et al. [117, 118]. Auxiliary pump for waste dialysate, heparin, electrolyte, and bicarbonate, sensors, and control system are not included

Fig. 7.31 Picture of the wearable artificial kidney described in [119]. Reproduced from the original paper, with permission

dialyzer and cartridges for dialysate regeneration containing urease, zirconium phosphate, hydroxyl zirconium oxide, and activated carbon. A dual-channel pulsatile blood pump drives both blood dialysate in countercurrent flow at a frequency of 110 cycles per minute, generating a transmembrane pressure difference up to 160 mmHg. These intermittent flow changes create a push–pull mechanism and increase the convective mass transfer across the dialyzer; in particular, an improvement in the removal of middle MW molecules was observed [118]. A volumetric pump removes a fixed amount of spent dialysate to provide controlled ultrafiltration; two other pumps are included to infuse anticoagulant in blood and a solution containing potassium, calcium, and magnesium to the regenerated dialysate. To ensure the wearability, the dialysate is continuously regenerated and its volume is reduced to 375 ml. While the

Fig. 7.32 β_2-microglobulin removal in different sorbent beds. Data from [118]: inlet concentration 548 µg/l, average flow rate 72.6 ml/min. In the device tested, four cartridges are present (in order along the dialysate flow: urease, zirconium phosphate, hydroxyl zirconium oxide, and activated carbon)

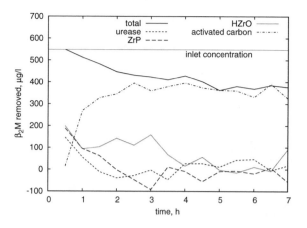

WAK uses the same sorbents as the REDY system, the operating pH is kept higher (7.4) in order to increase the zirconium phosphate capacity for ammonium ions; in vitro experiments showed that the amount of urea removed from the dialysate raises from 13.5 to 44 mg/g ZrP as pH increases from 5.5 to 7.4 [118]. The ability of the sorbent cartridge to remove β_2-microglobulin was also tested [118]: as reported in Fig. 7.32, β_2-microglobulin is mainly removed by activated carbon, even if in the first 1–2 h of treatment some adsorption onto the urease containing ZrP and HZrP layers occurs; after this time, the first three cartridges are saturated, but the activated carbon is still able to remove about 70 % of the β_2-microglobulin load with uniform operation and without saturation during the treatment.

A prototype of WAK (Xcorporeal Inc, Los Angeles, CA, USA) was tested in two clinical trials on 8 patients with end-stage kidney failure for a time from 4 to 8 h [119, 120]. The device is composed of a commercial 0.6 m^2 high-flux dialyzer (Gambro Dialysatoren, Hechingen, Germany), made of polysulfone, a pulsatile blood pump, powered by a 9 V battery. In the trials, blood and dialysate flow rates were set to about 60 and 50 ml/min, respectively. The membrane module used had an in vitro clearance of 50 ml/min for both urea and creatinine and 49 ml/min for phosphate and a sieving coefficient of 0.63 for β_2-microglobulin, with a dialysate flow of 500 ml/min. The amounts of urea, creatinine, phosphate, and β_2-microglobulin removed during the treatment are summarized in Table 7.15. During the study period, the clearances appeared to remain relatively stable and did not decrease significantly after 6–8 h. The blood and dialysate flow rates used were much lower than those in conventional dialysis, as the clearance of urea and creatinine; however, if the treatment can be carried out continuously, higher doses of dialysis can actually be delivered compared to standard protocols. Furthermore, the average β_2-microglobulin clearance is about 50 % of urea and creatinine clearance: this suggests that the pulsatile pump may produce a "push–pull" mechanism that improves β_2-microglobulin removal by convection and confirms that β_2-microglobulin is almost completely adsorbed by the sorbents. At the end of 2014, the WAK system was granted approval for human testing in the USA by FDA.

Table 7.15 Clearances of urea, creatinine, phosphate, and β_2-microglobulin during treatment with the wearable artificial kidney [119, 120]

	Amount removed, mg		Clearance, ml/min
	4 h	7–8 h	
Urea	7.0 ± 1.8	12.1 ± 5.1	22.7 ± 5.2
Creatinine	4.8 ± 1.0	9.5 ± 4.8	20.7 ± 5.0
Phosphate	300 ± 42	532 ± 400	21.7 ± 4.5
β_2-microglobulin	57 ± 26	125 ± 67	11.3 ± 2.3

The mean values of the amounts removed after 4 h are obtained from three patients, and those at 7–8 h are obtained from four patients

7.14 Bio-Artificial Kidneys

Renal replacement therapies based on artificial membrane devices are effective in changing the prognosis of renal failure and allowing prolonged survival of patients with final-stage renal disease; however, these therapies do not provide a complete replacement of the renal functions. Indeed, purely artificial devices provide only a more or less selective filtration function and are able to remove the end products of the metabolism, but they are unable to replace the regulatory, metabolic, and endocrine functions of healthy kidneys. In other words, the currently available artificial kidneys can be considered as substitutes for the glomerular functions, while the complex tubular functions require evolved specialized cells and cannot be replaced by an artificial membrane.

For this reason, efforts have been focused on the use tissue engineering to build a bio-artificial device, which provides the whole range of renal functions by including living tubular cells in association with a conventional membrane device (see [121] and [122] for a review).

A renal tubule cell-assist device (RAD) can be obtained by culturing renal tubule progenitor cells on hollow fiber membranes that provide the architectural scaffold for cell growth. Multifiber cartridges with membrane surface areas as large as $0.7\,\mathrm{m}^2$ and containing up to 10^8 cells have been realized and tested in vitro; active transport of sodium, bicarbonate, glucose, and organic anions, as well as production of ammonia and dihydroxyvitamin D_3, was proved; on the other hand, erythropoietin (which is synthesized from cells in the renal interstitium) was not produced in this device. On the basis of these results, an extracorporeal bio-artificial kidney was developed (see Fig. 7.33), including a standard hemofiltration cartridge in series with a renal tubule cell-assist device. In such device, the ultrafiltrate from the hemofiltration is delivered to the intraluminal compartment of the RAD, while blood is sent to the extraluminal compartment. The ultrafiltrate exiting the lumen of the RAD is discharged, while the blood from RAD is finally returned to the patient. The internal wall of the fibers is lined with living renal tubule cells that provide for the reabsorption of some valuable compounds and water and perform their metabolic and endocrine functions. The device thereby mimics the two-part function of the natural kidney:

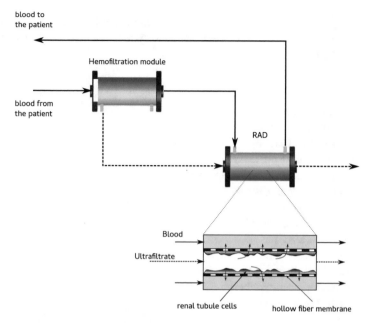

blood to
the patient

Hemofiltration module

blood from
the patient

RAD

Blood

Ultrafiltrate

renal tubule cells   hollow fiber membrane

Fig. 7.33 Schema of a circuit of a bio-artificial kidney (BAK), constituted of a hemofiltration cartridge and a renal tubule cell-assist device (RAD)

filtration of toxins followed by reclamation of salt, electrolytes, glucose, and water, plus the metabolic activity that only living cells can provide. The molecular weight cutoff of the hemofilter membrane provides immunoprotection of RAD cells, while the molecular weight cutoff of the membrane used as scaffold in RAD protects the patient from renal cell release [123]. The system was tested in FDA-approved clinical trials in the early 2000s, but in spite of the very encouraging results, development of RAD was discontinued, due to several problems related to cell supply, complex fabrication, and maintenance (maintenance and shipment to clinical site at 37 °C).

Recently, to overcome these problems, the bio-artificial renal epithelial cell system (BRECS) has been developed [124]. In BRECS, renal cells are cultured on porous solid disks (niobium-coated carbon disks) that adsorb matrix materials to promote cell growth and have favorable thermomechanical properties that enable cryopreservation. Indeed, the BRECS is an attached cell bioreactor, using primary epithelial cells seeded on porous disks; the disks are perfused with culture media containing oxygen and nutrients. When the optimal cell density is reached, the device can be cryopreserved (at −140 and −80 °C for long-term and short-term storages, respecitvely), and finally, thawed at 37 °C for therapeutic use. Even if BRECS has not yet been tested in clinical trials, it was designed for clinical perfusion with either ultrafiltered blood or body fluids (such as peritoneal fluid). The idea is that the ultrafiltrate from a standard hemodialyzer can be flown through BRECS; the conditioned ultrafiltrate, containing cellular therapeutic products, is then passed through an immunoisolation filter, which protects the patient from any large molecular by-products and/or cells released by the BRECS that could cause an adverse reaction, and finally returned to the patient blood.

Acronyms

AKI	Acute kidney injury
ARF	Acute renal failure
BAK	Bio-artificial kidney
CKD	Chronic kidney disease
CRRT	Continuous renal replacement therapy
ESRD	End-stage renal disease
HD	Hemodialysis
HF	Hemofiltration
HDF	Hemodiafiltration
MW	Molecular weight
RAD	Renal tubule cell-assist device
RRT	Renal replacement therapy
WAK	Wearable artificial kidney

Symbols

A	Membrane area
c	Concentration
$\bar{C}_s$	Mean solute concentration in the membrane
CL	Clearance
$\mathscr{D}$	Diffusivity
d_e	Equivalent diameter
DL	Dialysance
EKR	Equivalent renal clearance
G	Generation rate
Gz	Graetz number
GFR	Glomerular ultrafiltration rate
I	Protein intake rate
J_i	Solute i flux
J_v	Volume flux
k	Mass transport coefficient
k_E	Kinetic constant of the urease reaction
K_{GF}	Hydraulic permeability of the glomerular membrane
K_o	Overall mass transfer coefficient
K_r	Residual clearance
K_{UF}	Ultrafiltration coefficient
ℓ	Characteristic length
L	Membrane length
L_P	Hydraulic permeability
M_E	Urease mass

P	Pressure
Pe_{tm}	Transmembrane Peclet number
$\mathscr{P}$	Membrane permeability
Q	Volumetric flow rate
Q_F	Ultrafiltration rate
R	Ratio of the blood to dialysate flow rate
Re	Reynolds number
S	Passage area
$\mathscr{S}$	Sieving coefficient
Sc	Schmidt number
Sh	Sherwood number
TAC	Time-averaged concentration
v	Velocity
V	Volume
Z	Dimensionless, group
α_E	Dimensionless parameter defined in Eq. 7.48
δ	Membrane thickness
η_{reg}	Effectiveness of the regeneration system
ϕ	Fiber density
μ	Viscosity
π	Osmotic pressure
ρ	Density
σ	Reflection Staverman coefficient
ω	Diffusive permeation coefficient of the solute

Subscripts

1, 2	Phases 1, 2
B	Blood side
D	Dialysate side
dia	Intradialytic
i	Of component i
id	Interdialytic
in	Inlet stream
NP	Non-perfused
out	Outcoming
P	Perfused
pre	Predialysis

Superscripts

$'$	At the interface with the membrane

Chapter 8
Artificial and Bio-Artificial Liver

8.1 Introduction

The liver is a vital organ that performs a wide range of life-important functions and affects or controls almost every aspect of metabolism and most physiological regulatory processes; therefore, a severe loss of its function can often be life-threatening. Indeed, liver failure is associated with high plasmatic levels of various endogenous substances such as bilirubin, ammonia, protein breakdown products, and proinflammatory cytokines, which are involved in the aethiology of secondary multiorgan pathologies. Therefore, a first therapeutic approach to acute and acute-on-chronic liver failure is based on blood detoxification. Blood detoxification is successfully performed in the treatment of renal failure with hemodialysis, for the removal of small, water-soluble toxins such as urea and creatinine. However, toxins not cleared by the failing liver also include molecules that are tightly bound to plasma proteins, such as albumin, and more complex artificial devices, based on albumin dialysis, or plasma separation and regeneration, etc., have been developed for blood detoxification in liver failure.

On the other hand, it is evident that artificial devices for blood detoxification can at best represent a temporary solution aimed at keeping patients alive while waiting for either a recovery of liver functionality or an organ transplantation. In order to provide both detoxification and synthetic and regulatory functions of healthy liver, bio-artificial liver support systems, based on extracorporeal bioreactors with living liver cells, have been developed.

This chapter is mainly focused on the engineering aspects of the design of artificial liver support devices. After a short overview of liver functions and pathologies, aimed to define the minimum requirements for artificial devices, we report a survey of the different artificial liver support devices developed and used in clinical practice. Subsequently, starting from a description of the physicochemical phenomena that characterize each unit operation used in the detoxification process (albumin dialysis and toxin adsorption), mathematical models of dialysis and adsorption units are developed and combined into an LSD model. A proposal for a first approach to a

© Springer-Verlag London 2017
M.C. Annesini et al., *Artificial Organ Engineering*,
DOI 10.1007/978-1-4471-6443-2_8

patient device model is also considered in order to predict the evolution in time of toxin levels in patient blood and to evaluate the effectiveness of the treatment.

The chapter ends with a survey on the bio-artificial liver devices.

8.2 Structure and Function of Liver[1]

8.2.1 Liver Anatomy

Liver is a parenchymal organ devoted to many complex functions. It can be viewed as a small chemical plant where many physical and chemical operations are carried out. From an anatomic point of view, the liver is the largest gland of the body. The adult human liver weighs about 1.5 kg with sizes of 28 × 17 × 8 cm. It is located in the upper right quadrant of the abdomen, beneath the diaphragm. The liver consists of two major lobes (right and left) and two minor lobes (quadrate and caudate).

The major lobes are separated by the falciform ligament. The vascular system of the liver consists mainly of the hepatic artery, the portal vein, the hepatic veins, and the inferior vena cava (Fig. 8.1).

The bulk of the blood enters the liver through the hepatic artery (about 20 %) and the portal vein (about 80 %); the former supplies oxygen rich blood coming from the heart (about 1 l/min), while the latter supplies nutrient-rich blood coming from the digestive tract. After undergoing physical and chemical processes, the blood leaves the liver through the hepatic veins and the inferior vena cava.

In its secretory function, liver produces bile (an aqueous solution of salts, phospholipids, cholesterol, and bilirubin), which are drained through a system of ducts and conveyed to the gallbladder, where it is stored and concentrated before being discharged to the duodenum, when required by the organism. A scheme of the blood and bile circulation involving the liver is presented in Fig. 8.1. From a microscopic point of view, the liver is composed of hepatocytes (about 70 % vol.), which perform most of the metabolic functions, and Kupffer cells, which are the specialized macrophages involved in red blood cell breakdown. Hepatocytes are disposed in hexagonal structures, called liver *lobules*, around a central vein and crossed by a dense network of ducts where the cell surface is extended in order to increase the mass exchange surface.

[1]This and the following section were authored by G. Novelli V. Morabito, S. Novelli and S. Ruberto—Department "Paride Stefanini," General Surgery and Organ Transplantation, Extracorporeal Device and Artificial Liver Unit, University of Rome, La Sapienza.

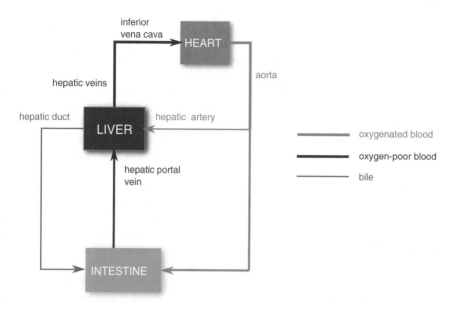

Fig. 8.1 Scheme of blood and bile circulation involving the liver

8.2.2   The Liver Functions

The main functions performed by the liver can be grouped in five categories:

1. Detoxification,
2. Secretion/excretion,
3. Synthesis,
4. Storage,
5. Immunological.

Detoxification processes involve capturing and neutralizing harmful substances, either coming from the gastrointestinal (GI) tract (e.g., drugs, poisons) or produced by the metabolic activity (e.g., ammonia, bilirubin, and alcohol). Detoxification from these substances can be carried out by reversible binding, chemical modification, and excretion or metabolic breakdown, especially for drugs and poisons. The reactions of the urea cycle, which convert ammonia to the much less toxic urea, are a typical example of the chemical detoxification processes occurring in the liver.

The excretory function is aimed at removing waste substances from blood, in order to maintain internal chemical homeostasis and to prevent damage to the organism. Among several important substances, liver produces bile, which is needed for fat absorption and for removing cholesterol and wastes. One of the main components of bile is bilirubin, a tetrapyrrole pigment produced in the reticuloendothelial cells as a product of heme catabolism, i.e., by the breakdown of old red blood cells. More in details, heme catabolism produces *unconjugated* (or *indirect*) biliribin, that is not soluble in water, but it is solubilized via tightly binding to albumin [125]. Unconjugated bilirubin is uptaken by hepatocytes, inside which it binds to several cytosolic proteins and, in healthy subjects, is conjugated with glucuronic acid to obtain the water-soluble *conjugated* (or *direct*) form. Much of the direct bilirubin goes in the bile and is expelled into the colon; a smaller part of conjugated bilirubin can be also recycled to the plasma and its renal excretion can be significant in some liver disease. Indeed, high hematic levels of bilirubin are generally a symptom of a liver pathology and bilirubin is commonly considered as an important marker to assess the status of the liver function; however, bilirubin is not the most dangerous among the substances whose hematic concentration increases in case of liver failure.

The synthetic function makes the liver similar to a chemical bioreactor where more than 500 different chemical reactions are carried out simultaneously. The liver synthesizes many biological compounds which play a key role in the health of the human body; among these are proteins, lipids, and carbohydrates. Almost all plasma proteins are synthetized in the liver. Fatty acids from the diet are used to form triglycerides, phospholipids, and cholesterol. Furthermore, insoluble lipids are converted to soluble forms. As for carbohydrates, glucose level is maintained stable by storing it as glycogen when the sugar level increases (for instance, after eating) and by decomposing glycogen into glucose and releasing it into blood when the sugar level decreases. In the case of chronic hepatitis, this function is impaired and blood sugar levels become unstable.

8.3   Liver Diseases and Artificial Support Systems

Many disorders and acute or chronic diseases can affect the liver with dangerous effects on the whole organism. Cysts, abscesses, and parasitic infections are considered more or less severe disorders. Likewise, viral hepatitis A, B, and C are inflammatory disorders, of which type B and, especially, C can be chronic. Acute liver failure (ALF) is characterized by the rapid loss of hepatocellular functions right after the appearance of the first signs of liver disease. Among the most common causes to this condition are the ingestion of a toxic substance, drug overdose, excessive consumption of alcohol, and viral hepatitis. Acute deterioration of liver function can occur also in patients with previously well-compensated chronic liver disease, either spontaneously or as a consequence of a precipitating event (acute on chronic liver failure, ACLF).

Among chronic liver diseases, cirrhosis is a very dangerous pathology characterized by the progressive replacement of liver tissue by fibrosis (scar tissue) and by the appearance of regenerative nodules. The cirrhotic liver gradually loses its functions. This serious and irreversible disease is often caused by alcoholism, hepatitis B or C. The medical treatment of cirrhosis is usually focused on preventing progression and complications such as hepatic encephalopathy and portal hypertension. Another very serious chronic liver disease is hepatocellular carcinoma (HCC) which has an incidence of 95 % among other primary liver tumors, such as hepatoblastoma, sarcoma, cholangiocarcinoma, and cystadenocarcinoma. One of the major causes of HCC is cirrhosis.

The liver has the unique capability to naturally regenerate the lost tissue; however, the rate of regeneration is too low to produce positive effects when more than 60 % of the liver function is lost. Therefore, it is necessary to substitute the failing liver with a healthy organ or support liver functions in waiting until the damage is repaired. Liver orthotopic transplantation is a surgical technique potentially useful when acute or chronic liver failure makes the organ irreversibly damaged, provided that other health conditions consent the operation. Unfortunately, the shortage of donors makes liver transplantation insufficient to cope with the large number of patients suffering from liver failure.

Therefore, in the last decades, different attempts to develop artificial liver support systems have been carried out, which led to the production and clinical testing of several devices. These devices are not intended for chronic use, but are aimed at bridging the patients to an organ transplantation or natural liver regeneration. Purely artificial devices provide for blood detoxification, which is the most urgent action to be undertaken in the treatment of liver failure; to that end, physical operations, mainly solid adsorption and membrane separation, are implemented.

More complex functions, such as synthesis, cannot be carried out with the currently available artificial devices. Therefore, scientific and technological efforts have also been devoted to develop a bio-artificial liver (BAL) consisting essentially of a bioreactor where hepatocytes are kept in a suitable environment for carrying out synthetic and metabolic function of the natural liver.

8.4 Artificial Liver Devices

Hemodialysis and related processes such as hemofiltration and hemodiafiltration are widely used for the therapy of kidney failure (see Chap. 6) when small or middle water-soluble toxins have to be removed from the blood. Unfortunately, the use of these therapies does not entail significant benefit in patients with liver failure. The main point is that in this case blood detoxification is more difficult than in kidney failure: as a matter of fact, most of the toxic substances that accumulate in hepatic failure, such as bile salts, bilirubin, aromatic amino acids, endotoxins and cytokines, are protein-bound toxins or have a high molecular weight. While hemodialysis or hemodiafiltration over rather selective membranes can remove small hydrophilic toxins, both effectively and selectively, they are unable to remove large protein-bound toxins; complex processes are required to cleanse the blood from these toxins, without depleting it of valuable macromolecules.

At present, different types of artificial liver devices are available, but all of them are based on extracorporeal albumin dialysis (ECAD), which is implemented in a single-pass mode or in a recirculating system, or a selective plasma separation, associated with albumin replacement or regeneration. In the following, the main artificial liver support devices are presented.

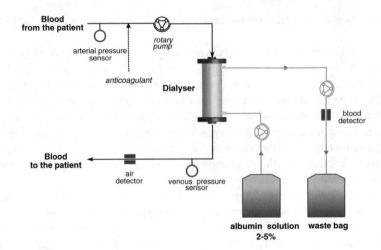

Fig. 8.2 Single-pass albumin dialysis

8.4.1   Single-Pass Dialysis (SPAD)

Single-pass albumin dialysis (SPAD) is the simplest form of albumin dialysis. Basically, this process (see Fig. 8.2) consists in blood dialysis across a special albumin-impregnated membrane against a concentrated albumin solution (albumin dialysate). The particular structure of the membrane and the presence of a binder in the dialysate allow for albumin-bound toxin transfer across the membrane, while the cutoff of the membrane is chosen so as to avoid transfer of albumin and higher molecular weight substances. SPAD can be implemented in a standard renal replacement therapy (RRT) machine, operating in the single-pass mode, i.e., with the discharge of the albumin-containing solution downstream of the dialyzer. Usually, the albumin concentration used in the dialysate is in the range 20–50 g/l (2–5 % by weight), while the dialysate flow rate is set to 1 l/h. Albumin in the dialysate binds lipophilic toxins, lowering their free concentration and enhancing their transfer through the membrane; thus, the higher the albumin concentration in the dialysate, the higher the effectiveness in removal of albumin-bound toxins, but the higher the costs of the treatment [126, 127].

8.4.2   Molecular Adsorbing Recirculating System (MARS)

A major drawback of SPAD is the high human albumin consumption and, therefore, cost of the treatment. Stange and Mitzner in 1993 [128, 129] developed a liver support system based on the albumin dialysis with online regeneration and recirculation of the albumin dialysate. This system, known as molecular adsorbing recirculating system or MARS® (Teraklin AG, Rostok, Germany now Gambro AB, Lund,

Sweden), applied for the first time in humans in 1996,[2] has become one of the most popular liver support system used in the clinical practice. As SPAD, MARS uses an albumin-impregnated polysulfone membrane and an albumin solution as dialyzer. In this case, however, the waste dialysate stream is not disposed, but cleaned and recycled to the dialyzer. In details, in the primary circuit, the patient's blood is pumped through a special high-flux dialyzer (MARSflux®) with a surface area of $2.1\,m^2$ and cutoff below 60 kDa; on the other side of the membrane, a dialysate solution (600 ml total volume, with a flow rate of 200–250 ml/min) with albumin concentration of 20 % is circulated in countercurrent flow. The spent albumin solution is sent to a secondary high-flux membrane module (diaFlux®) and undergoes a conventional dialysis to remove the small, water-soluble toxins. Afterward, the albumin solution is fed to two adsorption units, containing activated carbon (diaMars® AC250) and a cholestyramine anion-exchange resin (diaMars® E250), respectively. The use of two adsorption columns is aimed at removing a wider range of albumin-bound toxin types: activated carbon exhibits a high adsorption capacity for uncharged compounds, while the cholestyramine anion-exchange resin removes negatively charged compounds such as bile acids and bilirubin. The cleansed solution is recycled to the albumin dialyzer and is reused for toxin removal from the blood. Usually, one MARS session lasts 6–8 h, and after that time the albumin regeneration capacity of the adsorbers decreases significantly. MARS treatment results in the removal of water- soluble (ammonia, urea, and TNFα) and protein-bound substances (bilirubin, bile acids, tryptophan, short- and middle-chain fatty acids, and benzodiazepes). The clearances for strongly albumin-bound substances are between 10 and 60 ml/min; as for cytokines, significant removal has been measured, even if this does not always result in the decrease of blood cytokine levels [130–135].

8.4.3 Fractionated Plasma Separation and Adsorption (Prometheus)

In 2003, the Prometheus® artificial liver support device (Fresenius Medical Care AG, Bad Homburg, Germany), was introduced in the clinical practice. Prometheus is an implementation of the fractionated plasma separation and adsorption (FPSA) process proposed by Falkenhagen et al. [136] in 1999.

A scheme of the Prometheus system is presented in Fig. 8.3. The patient's blood is treated in a polysulfone dialyzer (AlbuFlow®) with a high molecular weight cutoff of the membrane (250 kDa), which allows the filtration of an albumin-rich plasma fraction (sieving coefficient 0.6), while blood cells and larger proteins are retained in the blood. The filtrate enters a secondary circuit where the albumin-rich plasma fraction is treated by adsorption to remove albumin-bound toxins. Two adsorption

---

[2]MARS® can be considered as the earliest application of the albumin dialysis process. However, here MARS® is presented after SPAD following the order of increasing complexity.

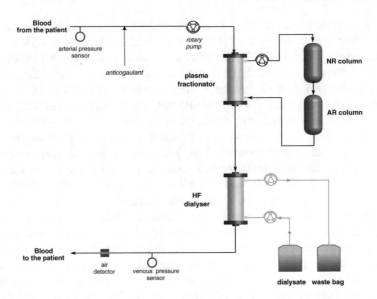

Fig. 8.3 Scheme of the Prometheus® system (NR neutral resin, AR anion-exchange resin)

columns, loaded with different solid adsorbents, are used: a neutral (Prometh® 01) and an anion-exchange (Prometh® 02) resin column. Downstream of the adsorption columns, the cleansed albumin solution is returned to the plasma fractionator, where it can be filtrated back in the primary circuit. High-flux hemodialysis is finally performed on the blood stream before returning it to the patient, in order to remove water-soluble toxins. No plasma is removed in the device and there is infusion of fresh (exogenous) albumin. Basically, FPSA can be considered as an ECAD process in which endogenous rather than exogenous albumin is used for the dialysate.

Even if appropriate, randomized, controlled, and adequately powered studies of artificial liver support systems are rare, it seems that Prometheus provides a more effective removal of tightly albumin-bound substances, such as unconjugated bilirubin than MARS, but lacks favorable hemodynamic effects [137–139]. Furthermore, Prometheus fails to reduce serum levels of cytokines, reportedly due to a low removal rate compared to the production rates [135, 140]. Neither Prometheus or MARS have the capacity to remove and replace albumin, nor it was shown that the removal of toxins by MARS restores functional capacity of albumin.[3]

---

[3]In contrast, high-volume plasma exchange, where patient's plasma was removed by plasma filtration and replaced by fresh frozen plasma, may be capable of removing and replacing albumin (and other plasma proteins).

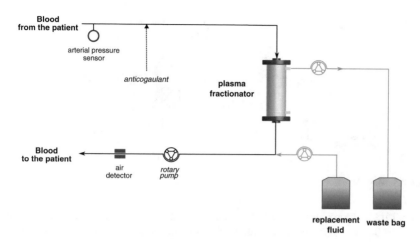

Blood from the patient

arterial pressure sensor

anticogaulant

plasma fractionator

Blood to the patient

air detector

rotary pump

replacement fluid   waste bag

Fig. 8.4 SEPET system

8.4.4 Selective Plasma Filtration Technology (SEPET Liver Assist Device)

A selective filtration of blood is also implemented in selective plasma filtration therapy (SEPET, Arbios Systems, Allendale, NJ, USA), which uses a membrane with a MW cutoff of 100 kDa.

A scheme of the Prometheus system is presented in Fig. 8.4. Similarly to FPSA, the membrane separates a plasma fraction containing several accumulated toxins (small molecular weight toxins and cytokines), while retaining important blood components, such as immunoglobulins, complement, protein, and growth factors. However, differently from FPSA, the filtrate is disposed and replaced by a combination of electrolyte solution, human albumin solution, and fresh frozen plasma. No data are yet available from human studies, although animal studies in ALF show improved survival and hepatic regeneration [141].

8.4.5 Hepa Wash

The Hepa Wash treatment (Hepa Wash GmbH, Monaco Munich, Germany) is a further variant of ECAD. Instead of albumin regeneration on adsorbent columns, this procedure uses changes in pH and temperature to allow albumin regeneration. In the Hepa Wash circuit (see Fig. 8.5), the albumin dialysate is divided into two parts. Each part undergoes a change of pH by addition of acid or base before passing through the filters; the change in the solution pH allows solubilization of ABTs, which can thus be removed by a filtration process. As an example, unconjugated bilirubin is virtually not soluble in water at neutral pH, but it is soluble in basic aqueous solutions. The

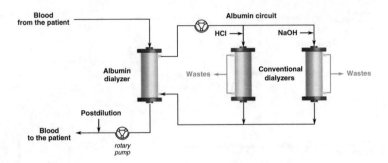

Fig. 8.5 Hepa Wash

two purified albumin streams are mixed and an albumin solution at physiological pH (range 6.9–7.6) is obtained and returned to the albumin dialyzer. This new process uses about 40 g of human serum albumin for each treatment. The current literature on this device concerns mainly the application to swines with acute liver failure. In a study of 2013 [142], the application of the Hepa Wash system proved safe and led to improved survival in pigs with acute liver failure. Before the use in clinical practice of this device, further tests are needed.

8.5 Modeling Toxin Removal in Liver Support Devices

8.5.1 Thermodynamics of Toxin Binding to Albumin

As previously described, liver failure results in the accumulation of both small water-soluble toxins and ABTs, such as middle-chain fatty acids, aromatic aminoacids, free phenols, and unconjugated bilirubin. While the former class of toxins can be easily removed by conventional dialysis, the main challenge faced by artificial liver support devices is to remove the toxins of the latter class without depleting blood of valuable molecules. The survey presented in Sect. 8.4 shows that unit operations such as membrane separation or adsorption are generally implemented in liver support devices. In such operations, chemical and phase equilibria of albumin–toxin systems play a major role and a proper understanding of these phenomena is required for a rational design of liver support systems.

For this reason, this section analyzes the thermodynamics of albumin–toxin binding and its implications on toxin phase equilibria. It is worth noting that, even if here explicit reference is made only to ABTs, the approach used here applies equally well to any other associating protein-toxin system.

The specific examples presented refer mainly to unconjugated bilirubin and tryptophan, which can be considered as significant representatives of two main classes of toxins involved in liver failure: strongly bound toxins with low water solubility (bilirubin) and low MW toxins that, though strongly bound to albumin, are present in plasma as free solutes in a non-negligible concentration (tryptophan).

8.5.1.1 Toxin-Albumin Binding

For the sake of the simplicity, only toxins (T) forming a 1:1 complex (AT) with albumin (A) will be considered. The binding reaction can be written as

$$A + T \rightleftarrows AT \tag{8.1}$$

The equilibrium constant of the above reaction is a

$$K_B = \frac{c_{AT}}{c_A c_T} \tag{8.2}$$

where c denotes the concentration and the subscripts refer to the components involved in the binding reaction (Eq. 8.1); clearly, the higher the binding constant, the higher the affinity for albumin and bound fraction of the toxin at equilibrium. Table 8.1 reports the values of the binding equilibrium constant for different albumin-bound toxins, including both endogenous and exogenous compounds. It is worth noting that bilirubin, which is usually considered as a standard marker of the clinical state of liver-failure patients, is very tightly bound to albumin, while tryptophan—an aromatic aminoacid involved in hepatic encephalopathy—shows a much lower affinity for albumin.

Table 8.1 Albumin–Toxin binding constants (*)

Ligand	K_B, M^{-1}	Reference
Bilirubin	$10^7 \div 10^8$	[143]
Tryptophan	10^4	[144]
Thyroxine	$10^5 \div 10^6$	[145, 146]
Bile acids	$10^4 \div 10^5$	[147, 148]
Phenol	3×10^4	[149]
Digoxin	10^5	[150]
Uracil	0.8×10^3	[151]
6n-Propyl-2-thiouracil	2.3×10^4	[151]
2-mercapto-1-methylimidazole	2.2×10^4	[151]
Indoxyl sulfate	$10^5 \div 10^6$	[152]
Salycilate	10^4	[153]
Acetylsalicylate	5×10^2	[153]
Benzylpenicillin	10^3	[153]
Ibuprofen	$10^4 \div 10^6$	[146]
Sulfobromophthalein	10^7	[146]
Diazepam	8×10^5	[154]
Cortisone	2×10^3	[146]

(*) dependence of the apparent binding constant with albumin concentration is reported for several toxins [155]

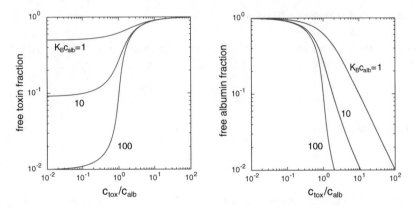

Fig. 8.6 Free fractions of toxin and albumin as a function of the toxin to albumin ratio for different binding constant values

The total (i.e., free and bound form) toxin and albumin concentrations are given as $c_{tox} = c_T + c_{AT}$ and $c_{alb} = c_A + c_{AT}$, respectively. The fraction of the free toxin $f_T = c_T/c_{tox}$ can be expressed as an implicit function of the albumin to toxin concentration ratio (c_{alb}/c_{tox}) and of the product $K_B c_{alb}$

$$K_B c_{alb} = \frac{1 - f_T}{f_T \left[1 - \dfrac{c_{tox}}{c_{alb}} (1 - f_T) \right]} \tag{8.3}$$

A similar expression, not reported for the sake of brevity, can be easily obtained for the free albumin fraction. Figure 8.6 reports the free fractions of toxin and albumin as a function of the albumin to toxin concentration ratio. For tightly bound toxins at concentration lower than albumin, i.e.,

$$K_B c_{alb} \gg 1; \quad c_{alb} > c_{tox} \tag{8.4}$$

equation 8.3 simplifies as

$$f_T = \frac{c_T}{c_{tox}} \simeq \frac{1}{1 + K_B c_{alb} \left(1 - \dfrac{c_{tox}}{c_{alb}} \right)} \tag{8.5}$$

Furthermore, if the toxin concentration is much lower than that of albumin, i.e.,

$$K_B c_{alb} \gg 1 \; ; \quad \frac{c_{tox}}{c_{alb}} \ll 1 \tag{8.6}$$

a further simplification of Eq. 8.5 is possible:

$$f_T \simeq \frac{1}{1 + K_B c_{alb}} \simeq \frac{1}{K_B c_{alb}} \tag{8.7}$$

It is worth noting that albumin concentration in blood is about 5×10^{-4} M (40 g/l); therefore, toxins with K_B of the order of magnitude of 10^6 M^{-1} or greater (like bilirubin) have only a small, if not negligible, unbound fraction in blood; other toxins, such as tryptofan, though strongly bound to albumin, are present in plasma also as free solutes in a non-negligible concentration.

8.5.1.2 Toxin Solubility

Binding with albumin results in a significant increase in plasma solubility for many toxins; this effect is dramatic for those toxins, such as bilirubin, that exhibit very low aqueous solubility in the free form. Indeed, one of the functions of plasmatic albumin is to carry and allow transport of insoluble compounds in blood, as in the case of toxin transport to the liver.

For a toxin having a free-form water solubility $c_{T,s}$, the solubility enhancement depends on the albumin concentration and binding constant as follows:

$$c_{tox,s} = c_{T,s} \left[1 + \frac{K_B c_{alb}}{1 + K_B c_{T,s}} \right] \tag{8.8}$$

where $c_{tox,s}$ is the apparent solubility of the toxin, which accounts for both the free and bound forms of the toxin. An example plot of the apparent solubility (saturation limit) of a toxin as a function of the albumin concentration is reported in Fig. 8.7. The values $c_{T,s} = 70$ nM and $K_B = 5 \times 10^6$ M^{-1} considered in the plot can be representative of the behavior of unconjugated bilirubin at normal albumin concentration [155, 156]; however, the presence on albumin of a second binding site for this toxin should be taken into account for a more accurate estimation of solubility.[4] It is important to highlight that in Fig. 8.7 the area above the curve corresponds to supersaturated solutions, where formation of aggregates and even precipitation of toxin may occur. As for unconjugated bilirubin, at a physiological albumin concentration of 450 μM, this occurs at about 100 μM of bilirubin, corresponding to a bilirubin to albumin

---

[4]It is generally agreed that the albumin molecule has one high-affinity binding site and one (or more) low-affinity site(s) for bilirubin; assuming independent binding of bilirubin to the two sites, the bound fraction of bilirubin is given by

$$1 - f_T = \frac{f_T K_{B1} c_{alb}}{1 + f_T K_{B1} c_{tox}} + \frac{f_T K_{B2} c_{alb}}{1 + f_T K_{B2} c_{tox}} \tag{8.9}$$

Furthermore, a decrease of the apparent binding constant K_B from 5×10^7 M^{-1} at low albumin concentration (15 μM) to 5.4×10^6 M^{-1} at higher albumin concentration (300 μM) is also observed [155]; therefore, the saturation limit is not linear in the albumin concentration [157].

Fig. 8.7 Saturation limit of toxin in an albumin solution. The plot refers to a toxin with a free-form saturation of 70 nM and $K_B = 5 \times 10^6\,\mathrm{M}^{-1}$

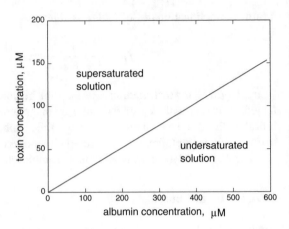

molar ratio of about 0.25, well below the saturation of the first albumin binding site; bilirubin aggregation in supersaturated solutions was proposed as a possible cause for bilirubin neurotoxicity [156, 157].

8.5.1.3   Phase Partition of Albumin-Bound Toxins

The partition equilibrium of an albumin-bound toxin between two aqueous phases (α and β) with different albumin concentration is considered. In this case, equilibrium conditions correspond to equal free toxin concentration in the two phases; therefore, the toxin partition coefficient ψ can be written as

$$\psi = \frac{c_{tox}^{\alpha}}{c_{tox}^{\beta}} = \frac{f_T^{\beta}}{f_T^{\alpha}} \tag{8.10}$$

where the superscripts α and β refer to the two liquid phases. If the conditions given in Eq. 8.6 hold, according to Eq. 8.7 the expression of the partition coefficient simplifies as

$$\psi = \frac{1 + K_B c_{alb}^{\alpha}}{1 + K_B c_{alb}^{\beta}} \simeq \frac{c_{alb}^{\alpha}}{c_{alb}^{\beta}} \tag{8.11}$$

and the equilibrium condition may be also rewritten in the following simpler form[5]

$$\frac{c_{alb}^{\alpha} - c_{tox}^{\alpha}}{c_{alb}^{\alpha}} = \frac{c_{alb}^{\beta} - c_{tox}^{\beta}}{c_{alb}^{\beta}} \quad \text{or} \quad \frac{c_{alb}^{\alpha} - c_{tox}^{\alpha}}{c_{tox}^{\alpha}} = \frac{c_{alb}^{\beta} - c_{tox}^{\beta}}{c_{tox}^{\beta}} \tag{8.12}$$

---

[5] When the affinity of the toxin for albumin is very high in both solutions (i.e., when the conditions given in Eq. 8.6 hold in both solutions), virtually all the toxin is in the bound form in both solutions and it may be assumed that $c_{AT} \simeq c_{tox}$ and $c_A \simeq c_{alb} - c_{tox}$).

Equation 8.12 shows that equal free albumin to total albumin ratios or free albumin to toxin ratios are reached in both phases.

This equilibrium condition holds for distribution of tightly bound toxins such as bilirubin between the solutions at different albumin concentration as well as for a dialysis process, with an albumin rejecting membrane and provide a rationale for some empirical findings reported in the literature [143, 158]

8.5.2 Dialysis of Albumin-Bound Toxins

Although their molecular weight is well below the cutoff of dialysis membranes, ABTs cannot be removed by the conventional dialysis processes used for water-soluble substances in RRT. Indeed, with the conventional dialysis mechanism, only free toxin molecules can cross the membrane, while the bound fraction is retained together with albumin; furthermore, due to the generally very low concentration of the free form of ABTs in aqueous solutions, the driving force for transmembrane flow of this species would be insufficient.

Starting from the first work of Stange and Mitzner in 1993 [129], albumin dialysis was introduced to address the above-described issues in the treatment of liver failure. Two fundamental ideas are at the basis of this process. Firstly, a binder is required in the dialysate in order to reduce the free toxin concentration and increase the driving force for toxin transmembrane flow [127]; albumin itself can be used as binder. Second, albumin molecules are able to exchange the bound toxin molecules over short distances; therefore, higher toxin transport rates can be obtained with a polymeric membrane impregnated with an albumin solution [159, 160]. A schematic representation of the transmembrane transport mechanism of albumin dialysis is reported in Fig. 8.8.

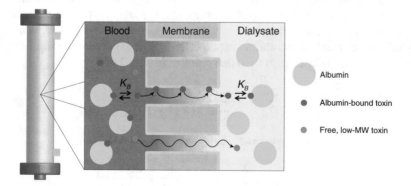

Fig. 8.8 Mechanisms of toxin permeation in albumin dialysis

By building on the model previously presented for kidney dialysis modules (see Sect. 7.8), the model of an albumin dialyzer can be developed including a suitable expression for ABT flux through the membrane. Since the driving force for the toxin flux through the membrane is the difference between the blood-side and dialysate-side free toxin concentration, the toxin flux is given by

$$J_{tox} = K_o \left(c_{T,B} - c_{T,D} \right) \tag{8.13}$$

where the mass transport coefficient K_o depends on the characteristics and albumin loading of the membrane.

Assuming that Eqs. 8.6 and 8.7 hold[6], the toxin flux may be also rewritten in terms of total toxin concentrations:

$$J_{tox} = K_o \left(\frac{c_{tox,B}}{1 + K_B c_{alb,B}} - \frac{c_{tox,D}}{1 + K_B c_{alb,D}} \right) \tag{8.14}$$

or as

$$J_{tox} = \frac{K_o}{1 + K_B c_{alb,B}} \left(c_{tox,B} - \psi c_{tox,D} \right) \tag{8.15}$$

It is worth noting that at very low toxin concentration ($c_{alb} \gg c_{tox}$), the driving force is a function of the difference between toxin to albumin molar ratios, in agreement with experimental results reported in the literature [158, 161].

Equation 8.15 may be coupled with mass balance equations in blood and dialysate to obtain the module dialysance,[7] which in this case is defined as

$$DL = Q_B \frac{c_{tox,B}^{in} - c_{tox,B}^{out}}{c_{tox,B}^{in} - c_{tox,B}^{out*}} \tag{8.16}$$

where Q_B is the blood flow rate and $c_{tox,B}^{out*}$ is the limiting toxin concentration that can be theoretically obtained in the outlet stream, i.e., the toxin concentration in blood at thermodynamic equilibrium with the inlet dialysate stream; superscripts *in* and *out* refer to inlet and outlet conditions, respectively. The expression obtained for the dialysance is perfectly analogous to that reported in Sect. 7.8:

$$\frac{DL}{Q_B} = \frac{1 - \exp\left[R\left(1 - Z\right)\right]}{Z - \exp\left[R\left(1 - Z\right)\right]} \tag{8.17}$$

but in this case, the dimensionless parameters R and Z are defined as

$$R = \frac{K_o A}{Q_B \left(1 + K_B c_{alb,B}\right)} = \frac{R_0}{1 + K_B c_{alb,B}} \tag{8.18}$$

---

[6] A rigorous development, without the assumption $c_{tox} \ll c_{alb}$ is reported in [143].
[7] For the derivation refer to Sect. 7.8.

$$Z = \frac{Q_B}{Q_D} \frac{1 + K_B c_{alb,B}}{1 + K_B c_{alb,D}} = Z_0 \frac{1 + K_B c_{alb,B}}{1 + K_B c_{alb,D}} \tag{8.19}$$

Equations 8.18 and 8.19 show that:

- for standard hemodialysis of unbound toxins, $R = R_0$ and $Z = Z_0$
- standard hemodialysis ($c_{alb,D} = 0$) of ABTs is largely less effective than for unbound toxins, due to both a decrease in R and $1/Z$ values, which is caused by the albumin concentration in the solution to be dialyzed. According to Meyer [162] the effectiveness should be improved only by increasing both R_0 and Q_D values;
- adding albumin as a toxin carrier to the dialysate is equivalent to increasing the dialysate flow rate; the higher the albumin concentration, the higher the dialysance that can be obtained [154]; nevertheless, a dialysance limiting value, DL_∞, is obtained for $1/Z \gg 1$

$$DL_\infty = Q_B[1 - \exp(-R)] \tag{8.20}$$

Such a limiting value is not affected by the albumin concentration in the dialysate and can be increased only with higher R_0 values; as a matter of fact, the membrane loading with albumin is aimed at exchanging toxins with albumin in the blood and to shuttle them across the membrane, thus improving R_0.

The order of magnitude of the product $K_o A$ for albumin dialyzers used in liver support devices can be estimated from experimental data reported by Stange and Mitzner [159], which measured the time course of bilirubin concentration in a two-compartment, closed-loop albumin dialysis system. Fitting of such experimental data with the model described in this section gives a $K_o A$ value of 2.6 ml/min; though this value is greater than for conventional dialysis of bilirubin, it is not comparable to those of uremic toxin in hemodialysis. With $K_o A$ values of this order of magnitude and typical operating conditions of albumin dialysis devices, a clearance of few percent points can be predicted. Indeed, in the clinical practice, the same flow rate (about 200 ml/min) is used.

8.5.3 Adsorption Units

Adsorption units are used in liver support devices to remove toxins directly from blood or plasma, such as in *haemoperfusion* and *plasmaperfusion* devices, or to regenerate the dialysate solution in recirculating devices like MARS®. In both cases, detoxification by adsorption is performed on an albumin-concentrated liquid phase and, depending on the adsorptive medium used, different classes of compounds can be removed, including both free and albumin-bound toxins. Indeed, several works are reported in the literature, which are aimed to evaluate the effectiveness of adsorption units for the removal of different toxins [163–166]; different adsorbent materials have been tested, including natural and synthetic adsorbents for bilirubin removal

[167–171]; furthermore, bio-affinity systems with adsorbents containing dyes or albumin as specific ligand for bilirubin have been also proposed [172, 173] to obtain a higher removal efficiency.

The most common adsorptive media used in artificial liver devices are activated carbons, neutral polymeric resins, and anion-exchange resins. Bilirubin adsorption onto activated carbon has been studied from serum or plasma [174–177]; some adsorption data on adsorption of aromatic aminoacids are also reported [177–179]. A neutral polymeric resin is used in Prometheus® to adsorb water-insoluble compounds, such as bile acids, phenols, and aromatic aminoacids. Anion-exchange resins (also used in MARS®) are suitable to remove negatively charged liver toxins, such as bilirubin.

Though some devices use fluidized and suspended-bed adsorption units [166, 177], fixed bed is by far the most common configuration used for adsorption columns in liver support devices.

As widely discussed in Chap. 4, which presents an overview of the main aspects related to adsorption from liquid solutions, the engineering analysis of an adsorption unit requires information about adsorption equilibrium and kinetics. Therefore, these two aspects are covered in the following sections, with specific focus on adsorption of ABTs.

8.5.3.1 Adsorption Equilibrium

ABTs, adsorption equilibrium is significantly affected by the presence of albumin concentration in the liquid phase; this is mainly due (but is not limited) to the competition between the toxin-albumin binding in the liquid phase and adsorption of the free toxin on the adsorbent surface [143, 180–183].

Expressions for an apparent ABT adsorption isotherms, in terms of total toxin concentration in the solution, are reported in the literature [181, 182, 184]; such expressions were derived by accounting for albumin–toxin association in the liquid phase and free toxin adsorption on the solid adsorbent according to the Langmuir isotherm (see Sect. 4.3):

$$n_{tox} = n_{max} \frac{f_T c_{tox}}{K_{lang} + f_T c_{tox}} \tag{8.21}$$

where n_{tox} is the toxin adborbed amount per unit sorbent mass; n_{max} and K_{lang} are the Langmuir isotherm parameters whose definition is given in Sect. 4.3. If albumin interaction with toxin adsorption is limited to toxin binding in the solution, n_{max} and K_{lang} should coincide with the Langmuir parameters of the free toxin adsorption isotherm in an albumin-free solution.

For tightly bound toxins and low toxin concentration (i.e., when the conditions given in Eq. 8.6 hold), the apparent isotherm can be put in a simple Langmuir-like form:

$$n_{tox} = \bar{n}_{max} \frac{c_{tox}}{\bar{K}_{lang} + c_{tox}} \tag{8.22}$$

with the following apparent parameters:

$$\bar{n}_{max} = \frac{n_{max}}{1 - K_{lang} K_B} \tag{8.23}$$

$$\bar{K}_{lang} = \frac{K_{lang} K_B}{1 - K_{lang} K_B} c_{alb} = m_{tox} c_{alb} \tag{8.24}$$

When $c_{tox} \ll K_{lang}$, Eqs. 8.21 and 8.22 can be replaced with the following linear approximations:

$$n_{tox} = \frac{n_{max}}{K_{lang}} f_T c_{tox} \tag{8.25}$$

$$n_{tox} = \frac{\bar{n}_{max}}{m_{tox} c_{alb}} c_{tox} \tag{8.26}$$

It is clear that, for a given total toxin concentration, different albumin concentrations correspond to different free toxin concentrations and, therefore, to different toxin adsorbed amounts. This is shown in the apparent adsorption isotherms presented above, which all predict a lower toxin adsorption as the concentration of albumin in the liquid solution increases. In Eq. 8.22, this effect is shown by the fact that $\bar{K}_{lang}$ is an increasing function of albumin concentration (see also Eq. 8.24).

According to the hypotheses described above, $\bar{n}_{max}$ should be independent of the albumin concentration in the solution (see Eq. 8.23). However, a dependence of this parameter on albumin concentration was observed when analyzing experimental data and attributed to possible non-chemical effects such as competitive adsorption or steric hindrance [181, 182], which are not accounted for in Eqs. 8.21 and 8.22.

Regardless of the phenomena actually involved, non-chemical effects can be accounted for by an empirical correction of the apparent adsorption isotherms. Table 8.2 reports a summary of the results obtained for bilirubin and tryptofan on different adsorptive media.

8.5.3.2 Fixed-Bed Adsorption

Modeling of fixed-bed adsorption columns is covered in Sect. 4.5. Here, the focus will be put on three possible operating regimes for the absorption column, which were previously referred to as *instantaneous*, *fast*, and *slow adsorption* regime, respectively.

In the istantaneous adsorption regime, the characteristic time of toxin transfer from the liquid to the solid phase is much shorter than the column residence time. In this case, the column behavior is characterized by a typical sigmoidal breakthrough curve and the toxin concentration in the column effluent is nearly zero until complete saturation of the solid phase is reached. In this regime, the saturation of the solid phase occurs progressively from the inlet to the outlet of the column, so that the

Table 8.2 Summary of adsorption isotherm parameters for bilirubin and tryptophan

Adsorbent material	Toxin	
	Bilirubin	Tryptofan
Activated carbon (05112, Fluka)	(Ref. Eq. 8.26) $\bar{n}_{max}/m_{tox} = 43.3\,\mu\text{mol/g}$	(Ref. Eq. 8.21) $n_{max} = n^0\left(1 - a\frac{c_{alb}/\bar{c}_{alb}}{1+c_{alb}/\bar{c}_{alb}}\right)$ $n^0 = 1330\,\mu\text{mol/g}$ $a = 0.644$ $\bar{c}_{alb} = 49.3\,\mu\text{M}$ $K_{lang} = 11.6\,\mu\text{M}$
Polymeric resin (Lewatit1064 MDPH, Bayer)	(Ref. Eq. 8.22) $\bar{n}_{max} = 0.482\,\mu\text{mol/g}$ $m_{tox} = 0.34$	
Anionic resin (IRA 400, Sigma Aldric)	(Ref. Eq. 8.22) $\bar{n}_{max} = a\exp(-c_{alb}/c^*)$ $a = 48.9\,\mu\text{mol/g}$ $c^* = 1245\,\mu\text{M}$ $m_{tox} = 0.24$	(Ref. Eq. 8.25) $n_{max}/K_{Lang} = 0.077\,\text{l/g}$

axial toxin concentration profile in the adsorbed phase is strongly non-uniform. If axial dispersion is neglected, the ideal column model applies to an adsorption unit operating in this regime, so that

$$\frac{c_{tox}^{out}(t)}{c_{tox}^{in}} = \begin{cases} 0 & t < t_{bt} \\ 1 & t > t_{bt} \end{cases} \;\; ; \;\; t_{bt} = \frac{Mn_{tox}^*}{Qc_{tox}^{in}} \tag{8.27}$$

where M the sorbent mass, Q the liquid phase volumetric flow rate, and n_{tox}^* the toxin adsorbed amount per unit sorbent mass in equilibrium with the feed solution. For the best detoxification performance, adsorption columns of liver support devices should be operated in this regime, in order to ensure complete toxin removal until saturation of the sorbent. Furthermore, in such regime, the only concern for the column design is to ensure that the time needed to saturate the solid phase, i.e., t_{bt}, is longer than the duration of a single treatment session. The column breakthrough time depends strongly on the solid adsorption capacity, which is one of the most important design parameters in this regime.

The ideal column model can also be used as an approximated description of the column behavior in the more common fast absorption regime, which, as an example, was observed for tryptofan fixed-bed adsorption on activated carbon from albumin-containing solutions [185].[8]

---

[8] With regards to the data presented in [185], it is interesting to note that the presence of albumin in the solution resulted in a "back translation" of the breakthrough curves and decrease of breakthrough time compared to tryptophan adsorption from albumin-free solutions. This effect increased with increasing albumin concentration, reflecting the reduction of tryptophan uptake from albumin-containing solutions.

In the slow adsorption regime, the toxin transfer characteristic time is longer than the column residence time. As a consequence, the toxin concentration in the column effluent is non-negligible even in the early operating-time of the column, when the solid phase is still far from saturation. This behavior has been observed with bilirubin adsorption from albumin-containing solutions [186] and also confirmed by some clinical data obtained during a MARS treatment [187]. In slow adsorption, the saturation of the solid phase proceeds in an rather uniform fashion throughout the column bed, so that the axial toxin concentration gradient in the adsorbed phase is small. Even if a more detailed description is possible (see Sect. 4.5 and [187]), in this case a much simpler model can be obtained by assuming that the composition of the solid phase is uniform and following a quasi-steady-state approach. Under these simplifications, the column behavior can be described by combining the steady-state toxin balance in the liquid phase

$$v\frac{dc_{tox}}{dz} + \frac{3}{R_p}(1-\varepsilon)k_c\left(c_{tox} - c_{tox}^*\right) = 0 \qquad (8.28)$$

with the transient toxin mass balance in the solid phase

$$Q\left(c_{tox}^{in} - c_{tox}^{out}\right) = M\frac{dn_{tox}}{dt} \qquad (8.29)$$

In Eqs. 8.28 and 8.29, c_{tox}^* is the local toxin concentration in the liquid phase at equilibrium with the solid phase, k_c the mass transfer coefficient, R_p the radius of adsorbent particles, v the liquid superficial velocity and ε is the bed void fraction.

For the sake of simplicity, a linear adsorption isotherm will be assumed for the analysis presented in this section, so that

$$n_{tox} = Kc_{tox}^* \qquad (8.30)$$

where, according to what reported in Sect. 8.5.3.1, the adsorption equilibrium constant K is a decreasing function of albumin concentration.

By substituting Eq. 8.30 into Eq. 8.29, a system of two ordinary differential equations is obtained, which can be analytically integrated with the following initial and boundary conditions

$$t = 0 \quad n_{tox} = 0; \qquad t > 0 \quad z = 0 \quad c_{tox} = c_{tox}^{in} \qquad (8.31)$$

to obtain

$$\frac{c_{tox}^{out}}{c_{tox}^{in}} = 1 - e^{-\frac{t}{\tau}}\left(1 - e^{-St}\right) \qquad (8.32)$$

where

$$St = \frac{3}{R_p}(1-\varepsilon)k_c\frac{V}{Q} \quad ; \quad \tau = \frac{KM}{Q}\frac{1}{1 - e^{-St}}$$

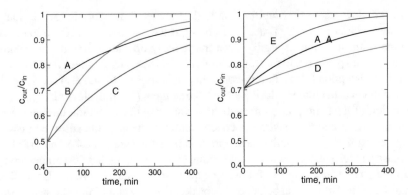

Fig. 8.9 Time evolution of the detoxification efficiency of the adsorption column for different conditions. The baseline case (curve A) refers to a solution with $c_{alb} = 100$ g/l, $St = St_A = 0.35$ and $\tau = \tau_A = 240$ min. The other curves refer to different sorbent mass, mass transfer coefficient, and albumin concentrations: (B) $k_c = 2k_{cA}$; (C) $M = 2M_A$; (D) $c_{alb} = 50$ g/l, (E) $c_{alb} = 200$ g/l

V being the column volume. The dimensionless number St gives a measure of the ratio of the toxin adsorption rate to the toxin axial transport rate; therefore, since Eq. 8.32 was derived assuming slow adsorption regime, its validity is limited to low values of St.

Equation 8.32 shows that the efficiency of the adsorption process has an initial, maximum value that depends strongly on St and then decays exponentially with time as equilibrium is approached. The time constant τ is proportional to K: it is clear that, the lower the adsorbed amount of toxin at equilibrium, the faster the column saturation. Again, it is worth noting that for albumin-bound toxins a decrease of K may be caused by an increase of albumin concentration.

In Fig. 8.9, Eq. 8.32 is plotted for different conditions. More specifically, we refer to a baseline case (curve A) with $St = St_A = 0.35$ and $\tau = \tau_A = 240$ min and we analyze the effect of an increase in the toxin adsorption rate (curve B) or an increase in the mass of adsorbent (curve C): it can be seen that an increase in the mass transfer rate is effective in reducing the toxin concentration at the beginning of the adsorption process, but the column saturates faster; on the other hand, an higher mass of adsorbent results both in a decrease of the initial outlet concentration and a slower saturation of the column. Curves D and E show the effect of the albumin concentration: the higher the albumin concentration, the faster the saturation of the column.

8.5.4 Albumin Dialysis and Regeneration

Artificial liver support devices are based on a single or a combination of detoxification units. In the previous sections, the units most commonly implemented in liver support devices were considered individually; here, the information presented for single units

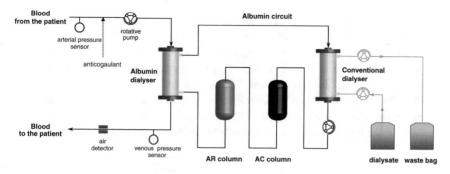

Fig. 8.10 Scheme of the MARS® system (AR anionic resin, AC activated carbon)

is combined to evaluate the overall efficiency of a liver support device. As an example, an albumin dialysis process with regeneration of the dialysate (as in MARS, see Fig. 8.10) will be considered.

To that end, it can be useful to evaluate the performance of each detoxification unit with respect to a given toxin in terms of the *detoxification efficiency,* η, which is defined as the fractional reduction of the toxin concentration that is achieved by the process

$$\eta = \frac{c_{tox}^{in} - c_{tox}^{out}}{c_{tox}^{in}} \tag{8.33}$$

From Eq. 8.33, it can be easily shown that the overall efficiency η_o of n detoxification processes in series is given by

$$\eta_o = 1 - \prod_{i=1}^{n}(1 - \eta_i) \tag{8.34}$$

where η_i is the detoxification efficiency of the ith process.

According to Eq. 8.34, the efficiency of the regeneration line (η_{reg}) of a recirculating ECAD device including one conventional dialysis module (CD), one activated carbon column (AC), and one anionic resin column (AR) in series is given by

$$\eta_{reg} = 1 - (1 - \eta_{dial,CD})(1 - \eta_{ads,AC})(1 - \eta_{ads,AR}) \tag{8.35}$$

with

$$\eta_{dial,CD} = \frac{CL}{Q_D} \tag{8.36}$$

and η_{ads} obtained from Eqs. 8.27 or 8.32.

The regeneration efficiency of each unit is different for each toxin to be removed. As for the elimination of hydrophilic, low molecular weight toxins, such as urea or creatinine, the most efficient unit is the secondary dialysis unit CD, while the adsorption columns play a minor role; therefore, in this case $\eta_{reg} \simeq \eta_{dial,CD}$ and the regeneration efficiency is almost time-independent. On the other hand, high-binding constant toxins are virtually not removed by conventional dialysis, so that the regeneration efficiency is determined by the adsorption columns and, therefore, is a decreasing function of time. Lower-binding constant toxins, such as tryptophan, are removed both in the conventional dialysis unit and in the adsorption columns.

The efficiency of the recirculating albumin dialysis process, η_{AD}, can be obtained by combining the dialysance of the albumin dialysis module (DL_{AD}) with the effectiveness of the regeneration circuit (details of derivation are reported in [188])

$$\frac{1}{\eta_{AD}} = \frac{Q_D}{DL_{AD}} + Z\frac{1 - \eta_{reg}}{\eta_{reg}} \tag{8.37}$$

Equation 8.37 shows how the efficiency of the recirculating albumin dialysis process depends both on membrane properties and regeneration line efficiency. For a high regeneration performance ($\eta_{reg} \to 1$), the overall system efficiency is controlled by the AD membrane properties, while, for a poor regeneration efficiency ($\eta_{reg} \to 0$), η_{AD} is strongly affected by the regeneration performance.

As an example of application of Eq. 8.37, it is possible to estimate the detoxification efficiency of the ECAD system with respect to bilirubin, which, as already pointed out, is commonly considered a reference toxin for the assessment of the performance of liver support devices. Because bilirubin is a strongly albumin-bound and negatively charged compound, its removal from the albumin dialysate occurs almost exclusively in the anionic resin column. Therefore, it may be assumed that $\eta_{dial,CD} \simeq 0$ and $\eta_{ads,AC} \simeq 0$. The results of the calculation are reported in Fig. 8.11. A baseline case (A) is considered, which corresponds to $c_{alb,D} = 100$ g/l, $St = 0.35$, $\tau = 120$ min and $R = 0.06$. Figure 8.11a shows that an improvement of the overall

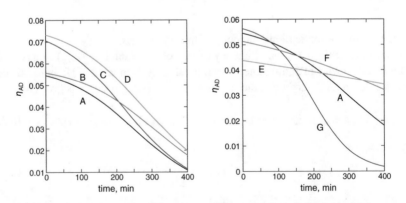

Fig. 8.11 Detoxification efficiency with respect to bilirubin, of a recirculating ECAD system. *Curve* A is a baseline case referring to $c_{alb,D} = 100$ g/l, $St = 0.35$, $\tau = 120$ min and $R = 0.06$. The other curves are obtained by changing the sorbent mass, membrane module dialysance and albumin concentration: (B) $M = 2M_A$; (C) $R = 2R_A$; (D) $M = 2M_A$ and $R = 2R_A$, (E)$c_{Alb,D} = 20$ g/l, (F) $c_{Alb} = 50$ g/l (G) $c_{Alb} = 200$ g/l

efficiency is obtained both by increasing the capacity of the AR column (curve B) or the dialysance of the membrane module (curve C). However, in the latter case, a better performance is obtained in the early treatment period, when adsorption columns are far from saturation and the membrane module performance controls the efficiency of the whole system; on the other hand, a higher bilirubin transfer from the blood to the dialysate increases the load on the adsorption column that saturates more rapidly and the overall efficiency decreases more rapidly. Obviously, an even higher efficiency can be obtained if the improvement of the membrane module dialysance is associated to an improvement of the AR capacity (curve D). Figure 8.11b shows the positive effect of albumin concentration in the dialysate during the early treatment period; on the other hand, if $c_{alb,D}$ is increased, column saturation occurs more rapidly and this causes a faster decay of the overall system clearance with time.

8.6 Bio-Artificial Devices

Purely artificial devices focus on the detoxifying function of the liver, the most urgent problem to face in the event of liver failure. Unfortunately, this kind of approach has shown to provide only very limited and temporary benefits for patients with ALF or with acute on chronic liver failure, waiting for either transplantation or liver regeneration. Removal of free and protein-bound toxins carried out in an artificial device seems to be insufficient, due to the lack of hepatocytes able to address more complex metabolic pathways; moreover, non-selective adsorption of substances from the patient's blood might have adverse effects on liver regeneration [189, 190]. In order to overcome this drawback, many researchers are attempting to develop bio-artificial livers (BALs) where a biological component has the role of performing the synthetic and metabolic functions of the natural liver.

8.6.1 BAL Systems

The various cutting-edge technologies proposed in literature differ both in the blood treatment circuit and in bioreactor design. A general scheme of a bio-artificial liver is reported in Fig. 8.12: plasma separated from the patient blood is sent to a reservoir and then pumped to an artificial detoxifying unit; the detoxified plasma is oxygenated and perfuses the bioreactor, where hepatocytes are cultured and kept viable at high cell density. The treated plasma is then recombined with the blood cells and pumped back into the patient's circulatory system.

The non-biological component includes a device for plasma treatment that can be based on charcoal hemoperfusion or on more sophisticated operations such as albumin dialysis used in the Molecular Adsorbent Recirculating System (MARS). Such a device is not always present since hepatocytes contained in the bioreactor could carry out detoxifying functions, but its presence is advisable in order to protect

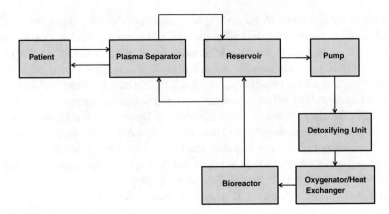

Fig. 8.12 Block diagram of BAL system

the cells from toxic injury. The core of a BAL is the bioreactor which must ensure the following main conditions [191]:

1. a medium suitable for growth and metabolism of the cells which should be cultured at high density and uniformly distributed in 3D configurations;
2. an efficient mass transport of nutrients toward the hepatocytes and of metabolic products from the cells to the plasma;
3. a system able to protect the hepatocytes population from the attack of the immune system;
4. a small dead space within the device.

In order to fulfill these requirements and to provide optimum conditions for cell viability and functions, various support systems, and bioreactor configurations have been proposed in literature. The main configurations implemented over the years are [191, 192] as follows:

- flat plate bioreactors where cells are cultured as monolayers;
- hollow fiber bioreactors where the bundle of small diameter hollow fibers divides the bioreactor in two compartments. Hepatocytes can be confined within or outside the fibers, whereas the medium or plasma flow in the other compartment;
- packed bed where cells are immobilized onto solid supports or scaffolds;
- fluidized-bed bioreactors containing hepatocytes entrapped in microcapsules suspended in the fluid phase;
- complex reactors based on braided bundles of fibers.

Support systems or scaffolds are required since hepatocytes need a solid surface for their adhesion and the formation of 3D structures. Without these conditions, hepatocytes undergo morphological and biochemical changes and quickly lose their functions. The central requirement in choosing and designing the bioreactor is to ensure all the cells are adequately supplied with oxygen and soluble nutrients and that the products of the hepatocytes metabolism can diffuse into the plasma. In other terms, mass transfer limitations represent a critical aspect in the performance of the BAL. Among the mass transport resistances, the external one can be reduced by increasing the medium flow rate, but the porous internal resistance is affected by the physical properties of the porous structure and by its thickness. Moreover, as shown in Chap. 5, concentration profiles of nutrients, metabolites and other essential compounds are greatly affected by reaction rates of cellular metabolic reactions. Unfortunately, as outlined by Catapano and Gerlach [193], only few kinetic equations are available today in literature and most of them refer to the oxygen consumption rate (OCR). They assume an independence of OCR from the concentration of oxygen and other compounds. On the contrary, some experiments have shown that OCR is affected, for instance, by ammonia concentration [194]. Another critical feature to be taken into account is the time-dependence of some properties, such as cell density distribution in 3D cultures. Therefore, the optimal design of the a BAL system can be carried out only by using a suitable mathematical model for simulating the operating conditions of the bioreactor. Some examples of dynamic models are given in [195, 196].

8.6.2 Bioreactors for BAL

Bioreactors used in BAL applications can be grouped in two main categories:

- Bioreactors based on 2D culture of cells adhering to scaffolds as fiber walls or flat plates;
- Bioreactors based on 3D cell structures obtained as clusters or entrapping hepatocytes into porous scaffolds or microcapsules.

Flat plates allow uniform cell distribution and microenvironments to be obtained but expose cells to sometimes unacceptable shear stresses and require very large devices because of the low surface area to volume ratio. In the attempt to limit the drawbacks due to shear stress and to protect the cells, microgrooves orthogonal to the flow direction have been engraved by photolithographic techniques onto the solid support [197]. Sometimes, sandwich configurations, with cells confined between two plates, have been used. Hollow fiber bioreactors, where nutrients perfuse through semipermeable membranes have been, up to now, the most used for clinical applications. In this kind of bioreactors, a set of hollow fiber capillaries is arranged in a shell-and-tube configuration (Fig. 8.13) where the membranes separate the blood plasma from the cells, allowing the transport of toxins and of oxygen and nutrients toward the cells and the transport of metabolic by-products from them to the plasma. The cells can be contained inside the capillaries or can adhere on the external surface of the membrane.

In the VitaGen ELADR system, hepatocytes adhere on the outer surface of hollow fiber while the blood plasma flows inside the lumen of the membrane. ELADR is a variant of Circe HepatAssist System (Fig. 8.14), which uses porcine hepatocytes whereas ELAD uses cells of human hepatoblastoma C3A line. The plasma is separated from the corpuscular part of the blood in a plasmapheresis device and sent by

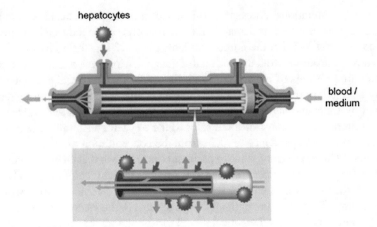

Fig. 8.13 Conceptual scheme of a hollow fiber bioreactor. Reproduced from [198], under the terms of the Creative Commons Attribution License

Fig. 8.14 Circe hepatAssist
system. Reproduced from
[200], with permission

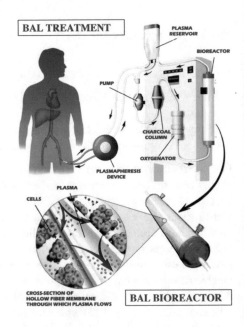

means of a pump onto a system of adsorption columns, devoted to the elimination of toxins, and then to an oxygenator before entering into the bioreactor. The pore size of the membrane is such as to prevent the passage of cells but to allow the passage of biomolecules to be exchanged between plasma and hepatocytes. In this configuration, plasma and nutrient medium flow in the same direction (cocurrent or countercurrent). An alternative possibility is a cross-flow scheme, with the medium flowing radially, from the outside to an inner core, and the cells growing inside the

Fig. 8.15 Steps followed for the entrapment of hepatocytes in the intraluminal space of a hollow fiber bio-artificial device: **a** Inoculation of hepatocytes in a collagen suspension; **b** gelation; **c** gel contraction and final flow configuration

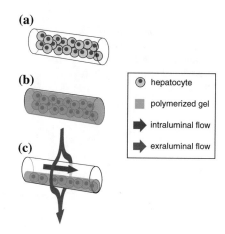

fibers. On the basis of a theoretical analysis, Adema and Sinskey [199] claim some possible advantages of this scheme on oxygen transport from the fiber wall to the cells and on mass transport from the bulk of the medium to the fiber wall.

In the LIVERx2000 system (Algenix Inc, USA), hepatocytes are contained inside the fibers and blood circulates outside the hollow fibers [201]. In the preliminary stage of inoculation, hepatocytes are suspended in a collagen solution and seeded into the lumen of the fibers (Fig. 8.15). The system is kept at 37 °C and after 15–20 min gelation of the solution gives a gel of polymerized collagen entrapping the cells. After 24 more hours, collagen gel contracts generating a layer containing the cells and a free volume available for intraluminal stream of medium or blood. When used as BAL, nutrient medium flows through the lumen of fibers. Entrapment of hepatocytes in a collagen bed (as in alginate gel) offers several positive features mainly enabled by the 3D reorganization of cells. In this 3D state, cells are stable for a longer time (weeks) with a viability higher than in the monolayer culture. Furthermore, metabolic functions appear to be improved. However, despite the close contact of the cells with the medium, the steep gradient of oxygen and other nutrient concentration across the cell layer sometimes results in cell necrosis at the innermost regions of the layer.

The transfer of oxygen and other nutrients to cells is improved in the Modular Extracorporeal Liver Support (MELS) system of the Charité (Universitätsmedizin Berlin, Germany) based on the hollow fiber bioreactor proposed and tested by Gerlach et al. [202]. This kind of bioreactor is based on a structure of four compartments arranged in a 3D network (Fig. 8.16) in the attempt to reproduce the liver vascular system.

The bioreactor is composed of several repeating units; each unit consists of two planar mats of hydrophilic membranes, with the direction of fibers in a mat orthogonal to the direction of the fibers in the underlying mat. The supply of oxygen and the removal of carbon dioxide is ensured by a third system of hydrophobic membranes interposed between the two mats. The fourth compartment is the volume outside the fibers used to culture the cells. In using this bioreactor as bio-artificial liver, a

Fig. 8.16 Scheme of the
four compartment perfused
bioreactor considered by
Gerlach et al. [202].
Reproduced from [203], with
permission

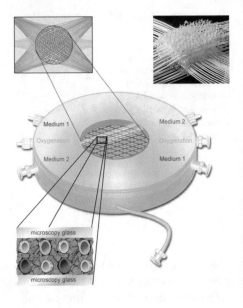

mat of hydrophilic membranes is used for medium perfusion whereas the other one
contains the plasma which flows countercurrent with the medium. Each mat in a
unit is connected with the corresponding mat of the subsequent unit, thus allowing
a quite easy scale-up of the system. The hollow fibers of the first and the second
compartment have separation and transport properties different from those of the third
compartment. Usually, porous hydrophilic polyethersulfone is used for medium and
plasma compartments whereas polypropylene is used for the hollow fibers devoted
to oxygenation and CO_2 removal.

Another system developed in Europe and clinically tested in a phase I safety trial
is the BAL of the Academic Medical Center (AMC) of the University of Amsterdam
(The Netherlands) [205]. The bioreactor is composed of a spirally wound polyester
rope and of a set of hydrophobic polypropylene hollow fiber membranes. The system
is contained in a polysulfone housing (Fig. 8.17). The hollow fibers act as spacers
between the layers of the spiral creating an extrafiber space. Hepatocytes are cultured
as small aggregates on the surface and inside the fibrous matrix of the rope (see
Fig. 8.15). Hollow fiber membranes are used for oxygen supply to the cells and
carbon dioxide removal. The medium circulates in the extrafiber space and is in direct
contact with hepatocytes. The AMC structure allows low-substrate and metabolite
gradients to be achieved because of the absence of a membrane separating medium
and cells; furthermore, the decentralized oxygenation system ensures an optimal
oxygen distribution to high-density cell aggregates. A comparison between MELS
cell module and AMC-BAL is reported in [206].

All solutions described so far are based on fixed beds of hepatocytes. Another
possible configuration for BAL is based on the fluidized-bed technology [207–209].
In this type of devices, liver cells are entrapped within biocompatible beads (such

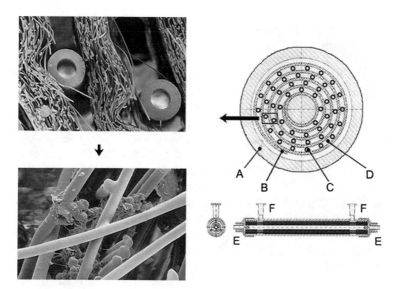

Fig. 8.17 AMC Bioreactor: scanning electron micrographs of the cell culture (*left*) and drawings of the axial and radial cross-sectional views of the bioreactor (*right*). A: bioreactor housing; B: support matrix for hepatocyte culture; C: hollow fibers (O₂); D: space for blood flow; E: gas inlet/outlet; F: blood inlet/outlet. Reproduced from [204], with permission

as alginate hydrogel) which are kept in continuous motion caused by the drag force of upward flow of medium and the downward force of gravity. Fluidization converts the bed of solid beads in an expanded mass that has many properties of a liquid. Due to the continuous motion, mass transfer between medium and solid particles is improved so that the intake of nutrients, oxygen, and lethal toxins and the release of essential proteins from cells to the fluid are favoured. Some studies on mass transfer in fluidized-bed bioreactor containing alginate beads are reported in the literature [210, 211]. Moreover, the activity of cells appears to be improved as well [212]. As a consequence, improvements in several clinical parameters, such as intracranial pressure, blood clotting, bilirubin conjugation, acidosis, and protein synthesis, have been observed [213]. However, special attention has to be devoted to transport properties of the material used for the beads and to the size of beads [214]. Poor transport properties or too large particles could give an insufficient concentration of oxygen and of other nutrients at the center of the particles with necrosis and death of the cells.

8.6.3 Cells Used in BAL

Whatever system is adopted, cells have to be extracted from an external source and immobilized on an adequate solid 3D support (scaffold) or entrapped in beads. Both human and porcine hepatocytes have been used in various proposed BALs. Other, genetically modified, cells have been used as well. The choice of pigs as source of liver cells is based on the analogy between the physiology of pigs and of humans as well as by the limited availability of human livers. Furthermore, porcine hepatocytes show a remarkable tendency to aggregation, which is a factor of utmost importance in order to implement the diversified functions of the liver. Liver-specific functions and cell viability show significant improvements when hepatocytes are cocultured with non-parenchymal cells. In particular, endothelial cells play an important role in the reorganization of cultured hepatocytes through the secretion of cytokines and nitrogen oxide (NO). Human cells, frequently used as the biological component of BALs, are cell lines coming from liver tumors. The lines most commonly used are HepG2 and HepaRG, both deriving from hepatocellular carcinoma. However, such cells contain a large amount of abnormal genetic component which can prevent the normal synthesis of proteins and the enzyme activity. A promising source of biological material is represented by differentiated hepatoblastoma cells, in particular HepG2/C3A line, artificially immortalized and modified by genetic engineering techniques in order to present the highly differentiated functions. Presently, however, culture of these cells have not shown a long term viability and furthermore exhibit a trend to malignant change. Anyway, the use of tumor or genetically modified cells involves the risk of leakage into the blood circulation system.

8.6.4   Cell Seeding

The first step of the process, leading to the biological construct to be used to mimic the liver tissue, is seeding of a large mass of liver cells on the scaffolds. Primary cells taken from liver tissue are cultured in Petri dishes or T-flasks and then harvested and seeded on synthetic scaffolds. Seeding operations must ensure a high density and uniform distribution over external and internal surface of the solid support. High initial density and uniformity of distribution promote the generation of a tissue suitable to perform liver functions. The seeding procedure used can considerably affect cell aggregation with the establishment of cellular polarity that plays a fundamental role in liver functions. However, distributing cells, uniformly and at high density, throughout the volume is not at all easy even with small pieces. In a commonly used method, cells are cultured in a flask before inoculation; subsequently, they are transported by gravity from the solution onto the surface of the scaffold and, eventually, into the porous structure. In this way low seeding efficiencies and non-uniform distribution are obtained. Better results are obtained by dynamic perfusion in which the solution containing the hepatocytes flows directly through the porous scaffold because of a pressure gradient. By using this procedure, a more uniform distribution is achieved in the seeding phase. The same technique can be used in the culture phase by flowing the nutrient medium through the porous structure. Commonly, the porosity of the scaffold is not uniform so that the smallest pores are not available for containing cells. Furthermore, cell adhesion to pore surface can hinder the penetration of further cells toward the innermost part of the solid. Moreover, biochemical and physical conditions should ensure cell viability and reorganization in a liver-like tissue. Nutrients and metabolites must diffuse through the structure and ensure everywhere adequate conditions for survival of hepatocytes. Because of concentration gradients linked to mass transport, however, different regions inside the construct could be characterized by different cell behaviors.

8.6.5 Scaffolds

An appropriate selection of the scaffold and of its properties is a fundamental topic in designing a bio-artificial liver. The geometric structure of the scaffold and its physical and chemical properties affect not only the capability to fix cells at high density, but also the metabolic behavior of the cells [193]. Many different porous materials and structures have been used in order to provide an adequate surface area for cell adhesion and the formation of 3-D structures. Three-dimensional structures allow a remarkably wider total surface than 2D structures so that a larger number of hepatocytes can be anchored. Furthermore, the high cell density obtainable in 3D structures stimulates the spontaneous cell aggregation and organization that is essential to liver-specific functions. Since 10–30 % of liver mass is indicated as the minimum to support life, a number of viable cells equivalent to this amount is the research target in the development of new bio-artificial devices. As for scaffold materials, polyethylene terephtalate (PET), polyvinyl alcohol (PVA), polyglycolic acid (PGA), and polylactic acid (PLA) are the most extensively used polymers, both in vitro and in vivo. Copolymers have been used as well. PET and PVA are non-degradable synthetic polymers that could set up a focus for infections and inflammations. On the contrary, PGA and PLA are degradable polymers with the advantage that they dissolve in time, leaving only the biological tissue when they are used as support of hepatocytes in transplantation of cells. Some of these materials are often provided with specific ligands on their surface to improve hepatocyte adhesion; for example, galactose ligand has been used with PET support. Besides the selection of material, the scaffold fabrication method is crucial to fulfill the requirements of a substrate for cell-based technologies. Recently, the electrospinning technique has proved to be a promising method in fabrication of polymeric scaffolds [215].

Acronyms

ABT	Albumin-bound toxin
ALF	Acute liver failure
ACLF	Acute-on-chronic liver failure
BAL	Bio-artificial liver
ECAD	Extracorporeal albumin dialysis
FPSA	Fractionated plasma separation and adsorption
GI	Gastro-intestinal

HCC	Hepatocellular carcinoma
MARS	Molecular adsorbent recirculating system
RRT	Renal replacement therapy
SEPET	Selective plasma filtration therapy
SPAD	Single-pass albumin dialysis

Symbols

A	Membrane area
CL	Membrane module clearance
c_A	Free albumin concentration
c_{AT}	Albumin–toxin complex concentration
$c_{alb,D}$	Albumin concentration in the dialysate solution
$c_{alb,B}$	Albumin concentration in blood
c_B^{in}	Toxin concentration at membrane module inlet (blood side)
c_B^{out}	Toxin concentration at membrane module outlet (blood side)
$c_B^{out,*}$	Minimum toxin concentration obtainable in the solution to purify (blood side)
c_D^{in}	Toxin concentration at membrane module inlet (dialysate side)
c_D^{out}	Toxin concentration at membrane module outlet (dialysate side)
c_T	Free toxin concentration
c_{tox}	Toxin concentration in the liquid phase
c_{tox}^*	Toxin concentration in the solid phase at equilibrium with the liquid phase
DL	Membrane module dialysance
H	Column height
J_v	Solvent flux across the membrane module
J_{tox}	toxin flux across the membrane module
K	Adsorption equilibrium constant
K_B	Albumin–toxin binding equilibrium constant
K_{lang}	Constant in Langmuir's adsorption isotherm
K_o	overall mass transfer coefficient
k_c	Mass transfer coefficient in the adsorption column
L	Membrane module length
n_{max}	Maximum toxin adsorbed amount
n_{tox}	Toxin adsorbed amount per unit sorbent mass
n_{tox}^*	Toxin adsorbed amount at the equilibrium with the feed solution
Q_B	Flow rate of solution to be purified
Q_D	Flow rate of cleansing solution
R	Sorbent particle radius
v	Liquid superficial velocity
ε	Adsorption column bed void fraction
ψ	Toxin partition coefficient
ρ	Sorbent intrinsic density
η_{dial}	Membrane module detoxification efficiency
η_{ads}	Adsorption column detoxification efficiency

η_{reg}	Regeneration circuit detoxification efficiency
η_{AD}	Recirculating albumin dialysis process efficiency
η_o	Overall efficiency
St	Stanton number
Q	Column feed flow rate
V	Column volume
M	Sorbent mass
t_{bt}	Column breakthrough time

References

1. R.B. Bird, W.E. Stewart, E.N. Lightfoot, *Transport Phenomena*, 2nd edn. (Wiley, New York, 2002)
2. T. Lango, T. Morland, A.O. Brubakk, Diffusion coefficients and solubility coefficients for gases in biological fluids and tissues: a review. Undersea Hyperb. Med. Soc. **23**, 247 (1996)
3. E.N. Lightfoot, K.A. Duca, The role f mass transfer in tissue function, in *The Biomedical Engineering Handbook: Second Edition*, chapter The role of Mass Transfer in Tissue Function, ed. by J.R. Bronzino (CRC Press LLC, Boca Raton, 2000)
4. B. Krause, M. Storr, T. Ertl, R. Buck, R. Deppish, H. Gohl, Polymeric membranes for biomedical applications. Chem. Ing. Tech. **75**, 1725–1732 (2003)
5. E.L. Cussler, *Diffusion: Mass Transfer in Fluid Systems* (Cambridge University Press, Cambridge, 2009)
6. S.I. Sandler, *Chemical, Biochemical and Engineering Thermodynamics* (Wiley, Hoboken, 2006)
7. J.M. Prausnitz, R.N. Lichtenthaler, E.G. de Azevedo, *Molecular Thermodynamics of Fluid-Phase Equilibria* (Pearson Education, Upper Saddle River, 1998)
8. S.R. de Groot, P. Mazur, *Non-equilibrium Thermodynamics* (Dover Publications Inc., New York, 1984)
9. J.A. Wesselingh, R. Krishna, *Mass Transfer in Multicomponent Mixtures* (Delft University Press, Delft, 2000)
10. W.R. Bowen, J.S. Welfoot, Modelling the performance of membrane nanofiltration: critical assessment and model development. Chem. Eng. Sci. **57**, 1121–1137 (2002)
11. J. Botella, P.M. Ghezzi, C. Sanz-Moreno, Adsorption in hemodialysis. Kidney Int. **58**, S-60–S-65 (2000)
12. J.F. Winchester, C. Ronco, J.A. Brady, E. Goldsa, J. Clemmer, L.D. Cowgill, T.E. Mullera, N.W. Levin, Sorbent augmented dialysis: minor addition of major advance in therapy? Blood Purif. **19**, 255–259 (2001)
13. A.L. Martinez de Francisco, P.M. Ghezzi, A. Brendolan, F. Fiorini, G. La Greca, C. Ronco, M. Arias, R. Gervasio, C. Tetta, Hemodiafilration with online regeneration of the ultrafiltrate. Kidney Int. **58**, S-66–S-71 (2000)
14. A.W. Adamson, *Physical Chemistry of Surfaces*, 4th edn. (Wiley, New York, 1982)
15. D.D. Do, *Adsorption Analysis: Equilibria and Kinetics* (Imperial College Press, London, 1998)
16. D.M. Ruthven, *Principles of Adsorption and Adsorption Processes* (Wiley, New York, 1984)
17. W.J. Thomas, B. Crittenden, *Adsorptionl Technology and Design* (Butterworth-Heinemann, Oxford, 1998)

© Springer-Verlag London 2017
M.C. Annesini et al., *Artificial Organ Engineering*,
DOI 10.1007/978-1-4471-6443-2

18. M.C. Annesini, C. Di Carlo, V. Piemonte, L. Turchett, Bilirubin and tryptophan adsorption in albumin-containing solutions: I. equilibrium isotherms on activated carbon. Biochem. Eng. J. **40**, 205–210 (2008a)

19. J. Lyklema, *Fundamentals of Interface and Colloid Science. Vol II: Solid-Liquid* (Academic Press, London, 1996)

20. I. Langmuir, The constitution and fundamental properties of solids and liquids. Part I. Solids. J. Am. Chem. Soc. **38**(11), 2221–2295 (1916)

21. C.J. Radke, J.M. Prausnitz, Thermodynamics of multi-solute adsorption from dilute liquid solutions. AIChE J. **18**(4), 761–768 (1972)

22. E. Glueckauf, J.J. Coates, Theory of chromatography. Part IV. The influences of incomplete equilibrium on the front boundary of chromatograms and on the effectiveness of separation. J. Chem. Soc. 1315–1321 (1947)

23. S. Sircar, R. Hufton, Intraparticle adsorbate concentration profile for linear driving force model. AIChE J. **46**(3), 659–660 (2000)

24. C.H. Liaw, J.S.P. Wang, R.A. Greenkorn, K.C. Chao, Kinetics of fixed-bed adsorption: a new solution. AIChE J. **25**(2), 376–381 (1979)

25. H.-K. Rhee, A. Rutherford, N.R. Amundson, *First-order Partial Differential Equations. Vol.1: Theory and Application of Single Equations* (Dover Publications Inc, New York, 2001)

26. P.V. Danckwerts, Continuous flow systems. Chem. Eng. Sci. **2**(1), 1–13 (1953)

27. J.F. Wehner, R.H. Wilhelm, Boundary conditions of flow reactor. Chem. Eng. Sci. **6**, 89–93 (1956)

28. L. Lapidus, N.L. Amundson, Mathematics of adsorption in beds. VI. The effect of longitudinal diffusion in ion exchange and chromatographic columns. J. Phys. Chem. **56**, 984–988 (1952)

29. H.S. Fogler, *Elements of Chemical Reaction Engineering*, Elements of Chemical Reaction Engineering (Prentice Hall PTR, 2006), https://books.google.it/books?id=QLt0QgAACAAJ. ISBN 9780130473943

30. J.E. Bailey, D.F. Ollis, *Biochemical Engineering Fundamentals*, Chemical engineering (McGraw-Hill, New York, 1986), https://books.google.it/books?id=zdBTAAAAMAAJ. ISBN 9780070032125

31. R. Leudeking, E.L. Piret, A kinetic study of the lactic acid fermentation. J. Biochem. Microbiol. Technol. Eng **1**, 393 (1959)

32. M. Cristina Annesini, G. Castelli, F. Conti, L. Conti De Virgiliis, L. Marrelli, A. Miccheli, E. Satori, Transport and consumption rate of o2 in alginate gel beads entrapping hepatocytes. Biotechnol. Lett. **22**(10), 865–870 (2000)

33. F. Meuwly, P.-A. Ruffieux, A. Kadouri, U. Von Stockar, Packed-bed bioreactors for mammalian cell culture: bioprocess and biomedical applications. Biotechnol. Adv. **25**(1), 45–56 (2007)

34. R. Langer, J.P. Vacanti, Tissue engineering. Science **260**(5110), 920–926 (1993)

35. J. Chaudhuri, M. Al-Rubeai, *Bioreactors for Tissue Engineering: Principles, Design and Operation*, Springer e-Books. Springer, SpringerLink (2006)

36. I. Martin, D. Wendt, M. Heberer, The role of bioreactors in tissue engineering. TRENDS Biotechnol. **22**(2), 80–86 (2004)

37. G.J. Peek, D. Elbourne, M. Mugford, R. Tiruvoipati, A. Wilson, E. Allen, F. Clemens, R. Firmin, P. Hardy, C. Hibbert, N. Jones, H. Killer, M. Thalanany, A. Truesdale, Randomised controlled trial and parallel economic evaluation of conventional ventilatory support versus extracorporeal membrane oxygenation for severe adult respiratory failure (cesar). Health Technol. Assess. **14**(35), 1–46 (2010). doi:10.3310/hta14350

38. T. Bein, F. Weber, A. Philipp, C. Prasser, M. Pfeifer, F.-X. Schmid, B. Butz, D. Birnbaum, K. Taeger, H.J. Schlitt, A new pumpless extracorporeal interventional lung assist in critical hypoxemia/hypercapnia. Crit. Care Med. **34**(5), 1372–1377 (2006). doi:10.1097/01.CCM. 0000215111.85483.BD

39. M.S. Hout, B.G. Hattler, W.J. Federspiel, Validation of a model for flow-dependent carbon dioxide exchange in artificial lungs. Artif. Organs **24**, 114–118 (2000)

40. M.W. Lim, The history of extracorporeal oxygenators. Anaesthesia **61**, 984–995 (2006). doi:10.1111/j.1365-2044.2006.04781.x

41. R.A. De Wall, The evolution of the helical reservoir pump-oxygenator system at the University of Minnesota. Ann. Thorac. Surg. **76**, 2210–2215 (2003)

42. C.N. Barnard, M.B. McKenzie, D.R. De Villiers, Preparation and assembly of the stainless steel sponge debubbler for use in the helix reservoir bubble oxygenator. Thorax **15**(3), 268–272 (1960)

43. Willem Kolff Interview - Academy of Achievement, November 1991 (2015), http://www.achievement.org/autodoc/page/kol0int-1. Accessed Feb 2015

44. R.J. Leonard, The transition from the bubble oxygenator to the microporous membrane oxygenator. Perfusion **18**, 179–183 (2003). doi:10.1191/0267659103pf659oa

45. J.D.S. Gaylor, Membrane oxygenators: current developments in design and applications. J. Biomed. Eng. **10**, 541–547 (1988)

46. H. Iwahashi, K. Yuri, Y. Nosé, Development of the oxygenator: past, present, and future. J. Artif. Organs **7**, 111–120 (2004). doi:10.1007/s10047-004-0268-6

47. M. Drummond, D.M. Braile, A.P.M. Lima-Oliveira, A.S. Camim, R.S. Kawasaki Oyama, G.H. Sandoval, Technological evolution of membrane oxygenators. Braz. J. Cardiovasc. Surg. **20**, 432–437 (2005)

48. D.F. Stamatialis, B.J. Papenburg, M. Gironés, S. Saiful, S.N.M. Bettahalli, S. Schmtmeier, M. Wessling, Medical applications of membranes: drug delivery, artificial organs and tissue engineering. J. Membr. Sci. **308**, 1–34 (2008)

49. G. Catapano, H.D. Papenfuss, A. Wodetzki, U. Baurmeister, Mass and momentum transport in extra-luminal flow (ELF) membrane devices for blood oxygenation. J. Membr. Sci. **184**, 123–135 (2001)

50. J.A. Potkay, The promise of microfluidic artificial lungs. Lab Chip **14**(21), 4122–4138 (2014)

51. G.S. Adair, A.V. Bock, H. Field Jr., The hemoglobin system vi. the oxygen dissociation curve of hemoglobin. J. Biol. Chem. **63**(2), 529–545 (1925)

52. J.H. Meldon, Blood-gas equilibria, kinetics and transport. Chem. Eng. Sci. **42**(2), 199–211 (1987)

53. F.B. Jensen, Red blood cell pH, the bohr effect, and other oxygenation-linked phenomena in blood o2 and co2 transport. Acta physiologica Scandinavica **182**(3), 215–227 (2004)

54. R.M. Winslow, M.L. Swenberg, R.L. Berger, R.I. Shrager, M. Luzzana, M. Samaja, L. Rossi-Bernardi, Oxygen equilibrium curve of normal human blood and its evaluation by adair's equation. J. Biol. Chem. **252**, 2331–2337 (1977)

55. A.V. Hill, The possible effects of the aggregation of the molecules of haemoglobin on its dissociation curves. J. Physiol. **40**, iv–vii (1910)

56. M. Samaja, A. Mosca, M. Luzzana, L. Rossi-Bernardi, R.M. Winslow, Equations and nomogram for the relationship of human blood p50 to 2,3-diphosphoglycerate, co2, and h+. Clin. Chem. **27**(11), 1856–1861 (1981)

57. M. Samaja, D. Melotti, E. Rovida, L. Rossi-Bernardi, Effect of temperature on the p50 value for human blood. Clin. Chem. **29**(1), 110–114 (1983)

58. C. Geers, G. Gros, Carbon dioxide transport and carbonic anydrase in blood and muscle. Physiol. Rev. **80**(2), 681–715 (2000)

59. P.M. Galletti, C.K. Colton, Artifcial lungs and blood-gas exchange devices, in *Biomedical Engineering Handbook*, vol. II, chapter 129, ed. by J.D. Bronzino (CRC Press LLC, Boca Raton, 1999)

60. S.R. Wickramasinghe, A.R. Goerke, J.D. Garcia, B. Han, Designing blood oxygenators. Ann. N.Y. Acad. Sci. **984**, 502–514 (2003)

61. S. Ranil Wickramasinghe, C.M. Kahr, B. Han, Mass transfer in blood oxygenators using blood analogue fluids. Biotechnol. Prog. **18**, 867–873 (2002)

62. N. Matsuda, K. Sakai, Blood flow and oxygen transfer rate of an outside blood flow membrane oxygenator. J. Membr. Sci. **170**, 153–158 (2000)

63. K. Nagase, F. Kohori, K. Sakai, Oxygen transfer performance of a membrane oxygenator composed of crossed and parallel hollow fibers. Biochem. Eng. J **24**, 105–113 (2005)

64. J.R. Zierenberg, H. Fujioka, K.E. Cook, J.B. Grotberg, Pulsatile flow and oxygen transport past cylindrical fiber arrays for an artificial lung: computational and experimental studies. J. Biomech. Eng. (2008)

65. T.J. Hewitt, B.G. Hattler, W.J. Federspiel, A mathematical model of gas exchange in an intravenous membrane oxygenator. Ann. Biomed. Eng. **26**, 166–178 (1998)

66. R.G. Svitek, W.J. Federspiel, A mathematical model to predict CO2 removal in hollow fiber membrane oxygenators. Ann. Biomed. Eng. **36**(6), 992–1003 (2008)

67. F. Yoshida, Prediction of oxygen transfer performance of blood oxygenators. Artif. Organs Today **2**, 237–252 (1993)

68. S.R. Wickramasinghe, B. Han, J.D. Garcia, R. Specht, Microporous membrane blood oxygenators. AIChE J. **51**, 656–670 (2005). doi:10.1002/aic.10327

69. M. Kaushik, M. Wojewodzka-Zelezniakowicz, D.N. Cruz, A. Ferrer-Nadal, C. Teixeira, E. Iglesias, J. Chul Kim, A. Braschi, P. Piccinni, C. Ronco, Extracorporeal carbon dioxide removal: the future of lung support lies in the history. Blood Purif. **34**(2), 94–106 (2012). doi:10.1159/000341904

70. S. Gramaticopolo, A. Chronopoulos, P. Piccinni, F. Nalesso, A. Brendolan, M. Zanella, D.N. Cruz, C. Ronco, Extracorporeal CO2 removal–a way to achieve ultraprotective mechanical ventilation and lung support: the missing piece of multiple organ support therapy, in *Cardiorenal Syndromes in Critical Care*, vol. 165, Contributions to Nephrology, ed. by C. Ronco, R. Bellomo, P.A. McCullough (S. Karger AG, 2010), pp. 174–184. doi:10.1159/000313757

71. M.E. Cove, G. MacLaren, W.J. Federspiel, J.A. Kellum, Bench to bedside review: extracorporeal carbon dioxide removal, past present and future. Crit. Care **16**(5), 232 (2012). doi:10.1186/cc11356

72. N.K. Burki, R.K. Mani, F.J.F. Herth, W. Schmidt, H. Teschler, F. Bonin, H. Becker, W.J. Randerath, S. Stieglitz, L. Hagmeyer, C. Priegnitz, M. Pfeifer, S.H. Blaas, C. Putensen, N. Theuerkauf, M. Quintel, O. Moerer, A novel extracorporeal CO2 removal system: results of a pilot study of hypercapnic respiratory failure in patients with copd. Chest **143**(3), 678–686 (2013). doi:10.1378/chest.12-0228

73. P.D. Wearden, W.J. Federspiel, S.W. Morley, M. Rosenberg, P.D. Bieniek, L.W. Lund, B.D. Ochs, Respiratory dialysis with an active-mixing extracorporeal carbon dioxide removal system in a chronic sheep study. Intensive Care Med. **38**(10), 1705–1711 (2012). doi:10.1007/s00134-012-2651-8

74. Heather Nolan, Dongfang Wang, Joseph B. Zwischenberger, Artificial lung basics: fundamental challenges, alternative designs and future innovations. Organogenesis **7**(1), 23–27 (2011)

75. S.D. Chambers, S.I. Merz, J.W. Mcgillicuddy, R.H. Bartlett, Development of the mc3 biolung, in *Engineering in Medicine and Biology, 2002. 24th Annual Conference and the Annual Fall Meeting of the Biomedical Engineering Society EMBS/BMES Conference, 2002. Proceedings of the Second Joint*, vol. 2 (IEEE, 2002), pp. 1581–1582

76. W.J. Federspiel, R.G. Svitek, *Lung, Artificial: Current Research and Future Directions*, chapter 150, pp. 922–931. doi:10.1081/E-EBBE-120022417

77. W. Johns, Blood/air mass exchange apparatus (2008)

78. J.D. Mortensen, G. Berry, Conceptual and design features of a practical, clinically effective intravenous mechanical blood oxygen/carbon dioxide exchange device (ivox). Int. J. Artif. Organs **12**(6), 384–389 (1989)

79. B.G. Hattler, G.D. Reeder, P.J. Sawzik, L.W. Lund, F.R. Walters, A.S. Shah, J. Rawleigh, J.S. Goode, M. Klain, H.S. Borovetz, Development of an intravenous membrane oxygenator: enhanced intravenous gas exchange through convective mixing of blood around hollow fiber membranes. Artif. Organs **18**(11), 806–812 (1994)

80. A.M. Guzmán, R.A. Escobar, C.H. Amon, Flow mixing enhancement from balloon pulsations in an intravenous oxygenator. J. Biomech. Eng. **127**(3), 400–415 (2005)

81. J.A. Potkay, A simple, closed-form, mathematical model for gas exchange in microchannel artificial lungs. Biomed. Microdevices **15**(3), 397–406 (2013)

82. C.D. Murray, The physiological principle of minimum work: I. the vascular system and the cost of blood volume. Proc. Natl. Acad. Sci. U.S.A. **12**(3), 207 (1926a)

83. C.D. Murray, The physiological principle of minimum work applied to the angle of branching of arteries. J. Gen. Physiol. **9**(6), 835–841 (1926b)

84. R. Sreenivasan, E.K. Bassett, D.M. Hoganson, J.P. Vacanti, K.K. Gleason, Ultra-thin, gas permeable free-standing and composite membranes for microfluidic lung assist devices. Biomaterials **32**(16), 3883–3889 (2011)

85. R.J. Gilbert, H. Park, M. Rasponi, A. Redaelli, B. Gellman, K.A. Dasse, T. Thorsen, Computational and functional evaluation of a microfluidic blood flow device. Asain J. **53**(4), 447–455 (2007)

86. P.W. Dierickx, F. De Somer, D.S. De Wachter, G. Van Nooten, P.R Verdonck. Hydrodynamic characteristics of artificial lungs. ASAIO J. **46**(5), 532–535 (2000)

87. G.A. Tanner, *Principles for clinical medicine*, chapter Kidney function (Lippincott Williams & Wilkins, Maryland, 2009), pp. 391–418

88. A.S. Levey, J. Coresh, E. Balk, A.T. Kausz, A. Levin, M.W. Steffes, R.J. Hogg, R.D. Perrone, J. Lau, G. Eknoyan, National kidney foundation practice guidelines for chronic kidney disease: Evaluation, classification, and stratification. Ann. Int. Med. **139**(2), 137–147 (2003). doi:10.7326/0003-4819-139-2-200307150-00013

89. A. Yavuz, C. Tetta, F.F. Ersoy, V. D'Intini, R. Ratanarat, M. De Cal, M. Bonello, V. Bordoni, G. Salvatori, E. Andrikos, G. Yakupoglu, N.W. Levin, C. Ronco, Uremic toxins: a new focus on an old subject. Semin. Dial. **18**(3), 203–211 (2005)

90. R. Vanholder, R. De Smet, G. Glorieux, A. Argiles, U. Baurmeister, P. Brunet, W. Clark, G. Cohen, P.P. De Deyn, R. Deppisch, B. Descamps-Latscha, T. Henle, A. Jorres, H.D. Lemke, Z.A. Massy, J. Passlick-Deetjen, M. Rodriguez, B. Stegmayr, P. Stenvinkel, C. Tetta, C. Wanner, W. Zidek. Review on uremic toxins: classification, concentration, and interindividual variability. Kidney Int. **63**(5), 1934–1943 (2003)

91. G. Eknoyan, The history of dialysis: the wonderful apparatus of john jacob abel called the "artificial kidney". Semin. Dial. **22**(3), 287–296 (2009)

92. W. Drukker, Haemodialysis: a historical review, in *Replacement of Renal Function by Dialysis*, ed. by J.F. Maher (Springer, Netherlands, 1989), pp. 20–86. ISBN 978-94-010-6979-3

93. T. Kapoian, R.A. Sherman, A brief history of vascular access for hemodialysis: an unfinished story. Semin. Nephrol. **17**(3), 239–245 (1997)

94. R. Mehta, Supportive therapies: intermittent hemodialysis, continuous renal replacement therapies, and peritoneal dialysis, in *Atlas of Disease of the Kidney*, chapter 19, ed. by R. Schrier (Current Medicine, Philadelphia, 1997), pp. 1–19

95. W.R. Clark, R.J. Hamburger, M.J. Lysaght, Effect of membrane composition and structure on solute removal and biocompatibility in hemodialysis. Kidney Int. **56**(6), 2005–2015 (1999a)

96. P.G. Kerr, L. Huang, Review: membranes for haemodialysis. Nephrology **15**(4), 381–385 (2010)

97. B. Krause, M. Storr, T. Ertl, R. Buck, H. Hildwein, R. Deppisch, H. Gohl, Polymeric membranes for medical applications. Chemie-Ingenieur-Technik, **75**(11), 1725–1732 (2003b)

98. J.R.L. Wolf, S. Zaltzman, Optimum geometry for artificial kidney dialyzers. Chem. Eng. Prog. Symp. Ser. **84**, 104–111 (1968)

99. L. Mi, S.-T. Hwang. Correlation of concentration polarization and hydrodynamic parameters in hollow fiber modules. J. Membr. Sci. **159**(1–2), 143–165 (1999)

100. M.-C. Yang, E.L. Cussler, Designing hollow-fiber contactors. AIChE J. **32**(11), 1910–1916 (1986)

101. L. Dahuron, E.L. Cussler, Protein extractions with hollow fibers. AIChE J. **34**(1), 130–136 (1988)

102. R. Prasad, K.K. Sirkar, Dispersion-free solvent extraction with microporous hollow-fiber modules. AIChE J. **34**(2), 177–188 (1988)

103. M.J. Costello, A.G. Fane, P.A. Hogan, R.W. Schofield, The effect of shell-side hydrodynamics on the performance of axial flow hollow fibre modules. J. Membr. Sci. **80**, 1–12 (1993)

104. S. Middleman, *Transport phenomena in the cardiovascular system*, Wiley-Interscience series on biomedical engineering (Wiley-Interscience, 1972). http://books.google.it/books?id=CttqAAAAMAAJ. ISBN 9780471602330

105. C. Cuccagna, Modello matematico per la valutazione delle prestazioni di un modulo a fibre cave per emodiafiltrazione. Master thesis in biomedical engineering, University of Rome La Sapienza, Engineering Faculty (2005)

106. J.A. Sargent, F.A. Gotch, *Principles and biophysics of dialysis, in Replacement of Renal Function by Dialysis* (Springer, Berlin, 1979), pp. 38–68

107. J.L. Walther, D.W. Bartlett, W. Chew, C.R. Robertson, T.H. Hostetter, T.W. Meyer, Downloadable computer models for renal replacement therapy. Kidney Int. **69**(6), 1056–1063 (2006)

108. S. Eloot, D. Schneditz, R. Vanholder, What can the dialysis physician learn from kinetic modelling beyond kt/vurea? Nephrol. Dial. Transpl. **27**(11), 4021–4029 (2012)

109. W.R. Clark, J.K. Leypoldt, L.W. Henderson, B.A. Mueller, M.K. Scott, E.F. Vonesh, Quantifying the effect of changes in the hemodialysis prescription on effective solute removal with a mathematical model. J. Am. Soc. Nephrol. **10**(3), 601–609, (1999b)

110. J.K. Leypoldt, B.L. Jaber, M.J. Lysaght, J.T. McCarthy, J. Moran. Kinetics and dosing predictions for daily haemafiltration. Nephrol. Dial. Transpl. **18**(4), 769–776 (2003)

111. F. Yoshida, Apparatus for treatment of artificial kidney dialyzing fluid, US Patent US4118314 A (1978)

112. J.W. Agar, Review: understanding sorbent dialysis system. Nephrology **15**, 406–411 (2010)

113. R.J. Wong, Cartridges useful in cleaning dialysis solutions (2006)

114. M. Roberts, The regenerative dialysis (redy) sorbent system. Nephrology **4**, 275–278 (1998)

115. S.R. Ash, The allient dialysis system. Semin. Dial. **17**, 164–166 (2004)

116. W.M. Kolff, B.H. Scribner, The development of renal hemodialysis. Nat. Med. **8** (2002)

117. V. Gura, C. Ronco, F. Nalesso, A. Brendolan, M. Beizai, C. Ezon, A. Davenport, E. Rambod, A wearable hemofilter for continuous ambulatory ultrafiltration. Kidney Int. **73**(4), 497–502 (2008)

118. V. Gura, A.S. Macy, M. Beizai, C. Ezon, T.A. Golper, Technical breakthroughs in the wearable artificial kidney (wak). Clin. J. Am. Soc. Nephrol. **4**(9), 1441–1448 (2009a)

119. A. Davenport, V. Gura, C. Ronco, M. Beizai, C. Ezon, E. Rambod, A wearable haemodialysis device for patients with end-stage renal failure: a pilot study. Lancet **370**, 2005–2010 (2007)

120. V. Gura, A. Davenport, M. Beizai, C. Ezon, C. Ronco, Beta2-microglobulin and phosphate clearances using a wearable artificial kidney: a pilot study. Am. J. Kidney Dis. **54**(1), 104–111 (2009b)

121. J. Jansen, M. Fedecostante, M.J. Wilmer, L.P. van den Heuvel, J.G. Hoenderop, R. Masereeuw, Biotechnological challenges of bioartificial kidney engineering. Biotechnol. Adv. **32**(7), 1317–1327 (2014), http://dx.doi.org/10.1016/j.biotechadv.2014.08.001, http://www.sciencedirect.com/science/article/pii/S0734975014001153. ISSN 0734-9750

122. H.D. Humes, D. Buffington, A.J. Westover, S. Roy, W.H. Fissell, The bioartificial kidney: current status and future promise. Pediatr. Nephrol. **29**(3), 343–351 (2014). doi:10.1007/s00467-13-2467-y. ISSN 0931-041X

123. D.A. Buffington, A.J. Westover, K.A. Johnston, H.D. Humes, The bioartificial kidney. Transl. Res. **163**(4), 342–351 (2014). http://dx.doi.org/10.1016/j.trsl.2013.10.006, http://www.sciencedirect.com/science/article/pii/S1931524413003782. Regenerative Medicine: The Hurdles and Hopes. ISSN 1931-5244

124. D.A. Buffington, C.J. Pino, L. Chen, A.J. Westover, G. Hageman, H. David Humes, Bioartificial renal epithelial cell system (brecs): a compact, cryopreservable extracorporeal renal replacement device. Cell. Med. **4**(1), 33–43 (2012). doi:10.3727/215517912X653328

125. D.G. Levitt, M.D. Levitt, Quantitative assessment of the multiple processes responsible for bilirubin homeostasis in health and disease. Clin. Exp. Gastroenterol. **7**, 307–328 (2014). doi:10.2147/CEG.S64283

126. P. Peszynski, S. Klammt, E. Peters, S. Mitzner, J. Stange, R. Schmidt, Albumin dialysis: single pass vs. recirculation (mars). Liver **22**(Suppl 2), 40–42 (2002)

127. T. Abe, M. Shono, T. Kodama, Y. Kita, M. Fukagawa, T. Akizawa, Extracorporeal albumin dialysis. Ther. Apher. Dial. **8**(3), 217–222 (2004). doi:10.1111/j.1526-0968.2004.00148.x

128. J. Stange, W. Ramlow, S. Mitzner, R. Schmidt, H. Klinkmann, Dialysis against a recycled albumin solution enables the removal of albumin-bound toxins. Artif. Organs **17**(9), 809–813 (1993a)

129. J. Stange, S. Mitzner, W. Ramlow, T. Gliesche, H. Hickstein, R. Schmidt, A new procedure for the removal of protein bound drugs and toxins. ASAIO J. **39**(3), M621–M625 (1993b)

130. S.R. Mitzner, S. Klammt, P. Peszynski, H. Hickstein, G. Korten, J. Stange, R. Schmidt, Improvement of multiple organ functions in hepatorenal syndrome during albumin dialysis with the molecular adsorbent recirculating system. Ther. Apher. **5**(5), 417–422 (2001)

131. S. Klammt, J. Stange, S.R. Mitzner, P. Peszynski, E. Peters, S. Liebe, Extracorporeal liver support by recirculating albumin dialysis: analysing the effect of the first clinically used generation of the marsystem. Liver **22**(Suppl 2), 30–34 (2002)

132. J. Stange, T.I. Hassanein, R. Mehta, S.R. Mitzner, R.H. Bartlett, The molecular adsorbents recycling system as a liver support system based on albumin dialysis: a summary of preclinical investigations, prospective, randomized, controlled clinical trial, and clinical experience from 19 centers. Artif. Organs **26**(2), 103–110 (2002)

133. J. Loock, S.R. Mitzner, E. Peters, R. Schmidt, J. Stange, Amino acid dysbalance in liver failure is favourably influenced by recirculating albumin dialysis (mars). Liver **22**(Suppl 2), 35–39 (2002)

134. S.R. Mitzner, Extracorporeal liver support-albumin dialysis with the molecular adsorbent recirculating system (mars). Ann. Hepatol. **10**(Suppl 1), S21–S28 (2011)

135. R.E. Stauber, P. Krisper, Mars and prometheus in acute-on-chronic liver failure: Toxin elimination and outcome [mars und prometheus beim acute-on-chronic leberversagen: Entfernung von toxinen und ergebnisse]. Transplantationsmedizin: Organ der Deutschen Transplantationsgesellschaft **22**(4), 333–338 (2010)

136. D. Falkenhagen, W. Strobl, G. Vogt, A. Schrefl, I. Linsberger, F.J. Gerner, M. Schoenhofen, Fractionated plasma separation and adsorption system: a novel system for blood purification to remove albumin bound substances. Artif. Organs **23**(1), 81–86 (1999)

137. K. Rifai, T. Ernst, U. Kretschmer, M.J. Bahr, A. Schneider, C. Hafer, H. Haller, M.P. Manns, D. Fliser, Prometheus-a new extracorporeal system for the treatment of liver failure. J. Hepatol. **39**(6), 984–990 (2003)

138. K. Rifai, M.P. Manns, Review article: clinical experience with prometheus. Ther. Apher. Dial. **10**(2), 132–137 (2006). doi:10.1111/j.1744-9987.2006.00354.x

139. K. Rifai, C. Tetta, C. Ronco, Prometheus: from legend to the real liver support therapy. Int. J. Artif. Organs **30**(10), 858–863 (2007)

140. P. Evenepoel, W. Laleman, A. Wilmer, K. Claes, D. Kuypers, B. Bammens, F. Nevens, Y. Vanrenterghem, Prometheus versus molecular adsorbents recirculating system: comparison of efficiency in two different liver detoxification devices. Artif. Organs, **30**(4), 276–284 (2006)

141. J. Rozga, Y. Umehara, A. Trofimenko, T. Sadahiro, A.A. Demetriou, A novel plasma filtration therapy for hepatic failure: preclinical studies. Ther. Apher. Dial. **10**(2), 138–144 (2006). doi:10.1111/j.1744-9987.2006.00355.x

142. A. Al-Chalabi, E. Matevossian, A.K. Thaden, P. Luppa, A. Neiss, T. Schuster, Z. Yang, C. Schreiber, P. Schimmel, E. Nairz, A. Perren, P. Radermacher, W. Huber, R.M. Schmid, B. Kreymann, Evaluation of the hepa wash® treatment in pigs with acute liver failure. BMC Gastroenterol. **13**, 83 (2013). doi:10.1186/1471-230X-13-83

143. M.C. Annesini, V. Piemonte, L. Turchetti, Albumin-bound toxin removal in liver support devices: case study of bilirubin adsorption and dialysis, in *Biochemical Engineering*, ed. by F.E. Dumont, J. Sacco (Nova Publisher, 2009a), pp. 321–339

144. R.H. McMEnamy, J.L. Oncley, The specific binding of l-tryptophan to serum albumin. J. Biol. Chem. **233**(6), 1436–1447 (1958)

145. T. Peters Jr., 3 - ligand binding by albumin, in *All About Albumin*, ed. by T. Peters (Academic Press, San Diego, 1995), pp. 76–132. http://dx.doi.org/10.1016/B978-012552110-9/50005-2. ISBN 978-0-12-552110-9

146. J. Steinhardt, J.A. Reynolds, Binding of neutral molecules, in *Multiple Equilibria in Proteins*, ed. by J. Steinhardt, J.A. Reynolds (Academic Press, 1969), pp. 84–175. http://dx.doi.org/10. 1016/B978-0-12-665450-9.50008-6. ISBN 978-0-12-665450-9

147. A. Roda, G. Cappelleri, R. Aldini, L. Barbara, Quantitative aspects of the interaction of bile acids with human serum albumin. J. Lipid Res. **23**(3), 490–495 (1982)

148. D. Rudman, F.E. Kendall, Bile acid content of human serum. ii. the binding of cholanic acids by human plasma proteins. J. Clin. Invest. **36**(4), 538–542 (1957)

149. S. Wada, S. Tomioka, I. Moriguchi, Protein bindings. vi. binding of phenols to bovine serum albumin. Chem. Pharm. Bull. (Tokyo) **17**(2), 320–323 (1969)

150. D.S. Lukas, A.G. De Martino, Binding of digitoxin and some related cardenolides to human plasma proteins. J. Clin. Invest. **48**(6), 1041–1053 (1969). doi:10.1172/JCI106060

151. A. Sulkowska, Interaction of drugs with bovine and human serum albumin. J. Mol. Struct. **614**(1–3), 227–232 (2002). doi:10.1016/S0022-2860(02)00256-9

152. E. Devine, D. Krieter, M. Rooth, J. Jankovski, H.-D. Lemke, Binding affinity and capacity for the uremic toxin indoxyl sulfate. Toxins **6**(2), 416–430 (2014). doi:10.3390/toxins6020416. ISSN 2072-6651

153. R. Brodersen, Competitive binding of bilirubin and drugs to human serum albumin studied by enzymatic oxidation. J. Clin. Invest. **54**(6), 1353–1364 (1974). doi:10.1172/JCI107882

154. K. Drexler, C. Baustian, G. Richter, J. Ludwig, W. Ramlow, S. Mitzner, Albumin dialysis molecular adsorbents recirculating system: impact of dialysate albumin concentration on detoxification efficacy. Ther. Apher. Dial. **13**(5), 393–398 (2009). doi:10.1111/j.1744-9987. 2009.00757.x

155. R.A. Weisiger, J.D. Ostrow, R.K. Koehler, C.C. Webster, P. Mukerjee, L. Pascolo, C. Tiribelli, Affinity of human serum albumin for bilirubin varies with albumin concentration and buffer composition: results of a novel ultrafiltration method. J. Biol. Chem. **276**(32), 29953–29960 (2001). doi:10.1074/jbc.M104628200

156. J.D. Ostrow, P. Mukerjee, C. Tiribelli, Structure and binding of unconjugated bilirubin: relevance for physiological and pathophysiological function. J. Lipid. Res. **35**(10), 1715–1737 (1994)

157. J.D. Ostrow, L. Pascolo, C. Tiribelli, Reassessment of the unbound concentrations of unconjugated bilirubin in relation to neurotoxicity in vitro. Pediatr. Res. **54**(6), 926 (2003). doi:10. 1203/01.PDR.0000103388.01854.91

158. C. Steiner, S. Sen, J. Stange, R. Williams, R. Jalan, Binding of bilirubin and bromosulphthalein to albumin: implications for understanding the pathophysiology of liver failure and its management. Liver Transpl. **10**(12), 1531–1538 (2004). doi:10.1002/lt.20323

159. J. Stange, S. Mitzner, A carrier-mediated transport of toxins in a hybrid membrane. Safety barrier between a patients blood and a bioartificial liver. Int. J. Artif. Organs **19**(11), 677–691 (1996)

160. D. Ge, W. Dewang, W. Shi, Y. Ma, X. Tian, P. Liang, Q. Zhang, An albumin-fixed membrane for the removal of protein-bound toxins. Biomed. Mater. **1**(3), 170–174 (2006). doi:10.1088/ 1748-6041/1/3/012

161. N. Dammeir, A. Baumann, M. Suraj, R. Weiss, S. Mitzner, R. Schmidt, J. Stange, Capacity of different adsorbent combinations to maintain a high toxin/albumin gradient during albumin dialysis, in *International Symposium on Albumin Dialysis in Liver Disease. Book of Abstracts 2002-2008* (2008)

162. T.W. Meyer, E.C. Leeper, D.W. Bartlett, T.A. Depner, Y.Z. Lit, C.R. Robertson, T.H. Hostetter, Increasing dialysate flow and dialyzer mass transfer area coefficient to increase the clearance of protein-bound solutes. J. Am. Soc. Nephrol. **15**(7), 1927–1935 (2004)

163. P. Evenepoel, B. Maes, A. Wilmer, F. Nevens, J. Fevery, D. Kuypers, B. Bammens, Y. Vanrenterghem, Detoxifying capacity and kinetics of the molecular adsorbent recycling system. Contribution of the different inbuilt filters. Blood Purif. **21**(3), 244–252 (2003). doi:10.1159/ 000070697

164. D. Falkenhagen, M. Brandl, J. Hartmann, K.-H. Kellner, T. Posnicek, V. Weber, Fluidized bed adsorbent systems for extracorporeal liver support. Ther. Apher. Dial. **10**(2), 154–159 (2006). doi:10.1111/j.1744-9987.2006.00357.x

165. J. Steczko, S.R. Ash, D.E. Blake, D.J. Carr, R.H. Bosley, Cytokines and endotoxin removal by sorbents and its application in push-pull sorbent-based pheresis: the biologic-dtpf system. Artif. Organs **23**(4), 310–318 (1999)

166. A.T. Peter, S.R. Ash, J. Steczko, J.J. Turek, D.E. Blake, D.J. Carr, W.R. Knab, R.H. Bosley, Push-pull sorbent-based pheresis treatment in an experimental canine endotoxemia model: preliminary report. Int. J. Artif. Organs **22**(3), 177–188 (1999)

167. C.R. Davies, P.S. Malchesky, G.M. Saidel, Temperature and albumin effects on adsorption of bilirubin from standard solution using anion-exchange resin. Artif. Organs **14**(1), 14–19 (1990)

168. S. Nakaji, N. Hayashi, Bilirubin adsorption column medisorba bl-300. Ther. Apher. Dial. **7**(1), 98–103 (2003)

169. T. Chandy, C.P. Sharma, Polylysine-immobilized chitosan beads as adsorbents for bilirubin. Artif. Organs **16**(6), 568–576 (1992)

170. J.A. Costanzo, C.A. Ober, R. Black, G. Carta, E.J. Fernandez, Evaluation of polymer matrices for an adsorptive approach to plasma detoxification. Biomaterials **31**(10), 2857–2865 (2010). doi:10.1016/j.biomaterials.2009.12.036

171. A.V. Nikolaev, Y.A. Rozhilo, T.K. Starozhilova, V.V. Sarnatskaya, L.A. Yushko, S.V. Mikhailovskii, A.S. Kholodov, A.I. Lobanov, Mathematical model of binding of albumin-bilirubin complex to the surface of carbon pyropolymer. Bull. Exp. Biol. Med. **140**(3), 365–369 (2005)

172. A. Denizli, M. Kocakulak, E. Piskin, Bilirubin removal from human plasma in a packed-bed column system with dye-affinity microbeads. J. Chromatogr. B. Biomed. Sci. Appl. **707**(1–2), 25–31 (1998)

173. C. Alvarez, M. Strumia, H. Bertorello, Synthesis and characterization of a biospecific adsorbent containing bovine serum albumin as a ligand and its use for bilirubin retention. J. Biochem. Biophys. Methods **49**(1–3), 649–656 (2001)

174. E.H. Dunlop, R.D. Hughes, R. Williams, Physico-chemical aspects of the removal of protein-bound substances by charcoal and other adsorbents of potential value in systems of artificial liver support: Part i-equilibrium properties. Med. Biol. Eng. Comput. **16**(4), 343–349 (1978)

175. V.G. Nikolaev, V.V. Sarnatskaya, V.L. Sigal, V.N. Klevtsov, K.E. Makhorin, L.A. Yushko, High-porosity activated carbons for bilirubin removal. Int. J. Artif. Organs **14**(3), 179–185 (1991)

176. V.V. Sarnatskaya, W. Edward Lindup, P. Walther, V.N. Maslenny, L.A. Yushko, A.S. Sidorenko, A.V. Nikolaev, V.G. Nikolaev, Albumin, bilirubin, and activated carbon: new edges of an old triangle. Artif. Cells Blood Substit. Immobil. Biotechnol. **30**(2), 113–126 (2002)

177. S.R. Ash, T.A. Sullivan, D.J. Carr, Sorbent suspensions vs. sorbent columns for extracorporeal detoxification in hepatic failure. Ther. Apher. Dial. **10**(2), 145–153 (2006). doi:10.1111/j.1744-9987.2006.00356.x

178. M.A. Tsvetnov, V.V. Khabalov, N.B. Kondrikov, Sorption of amino acids from aqueous solutions by a polarized carbon adsorbent. Colloid J. **63**(2), 248–252 (2001)

179. S. Mitzner, S. Klammt, J. Stange, R. Schmidt, Albumin regeneration in liver support-comparison of different methods. Ther. Apher. Dial. **10**(2), 108–117 (2006). doi:10.1111/j.1744-9987.2006.00351.x

180. M.C. Annesini, L. Di Paola, L. Marrelli, V. Piemonte, L. Turchetti, Bilirubin removal from albumin-containing solution by adsorption on polymer resin. Int. J. Artif. Organs **28**(7), 686–693 (2005a)

181. M.C. Annesini, C. Di Carlo, V. Piemonte, L. Turchetti, Bilirubin and tryptophan adsorption in albumin-containing solutions. i. equilibrium isotherms on activated carbon. Biochem. Eng. J. **40**(2), 205–210 (2008b). doi:10.1016/j.bej.2007.12.010

182. M.C. Annesini, V. Piemonte, L. Turchetti, Adsorption of albumin-bound toxins on anion exchange resin: an equilibrium study. Asia-Pac. J. Chem. Eng. **7**(4), 510–516 (2012). doi:10.1002/apj.600

183. J.F. Patzer, 2nd. Thermodynamic considerations in solid adsorption of bound solutes for patient support in liver failure. Artif. Organs **32**(7), 499–508 (2008). doi:10.1111/j.1525-1594.2008.00581.x

184. M.C. Annesini, L. Di Paola, L. Marrelli, V. Piemonte, L. Turchetti, Bilirubin removal from albumin-containing solution by adsorption on polymer resin. Int. J. Artif. Organs **28**(7), 686–693 (2005b)

185. M.C. Annesini, V. Piemonte, L. Turchetti, Removal of albumin-bound toxins from albumin-containing solutions: tryptophan fixed-bed adsorption on activated carbon. Chem. Eng. Res. Des. **88**(8), 1018–1023 (2010). doi:10.1016/j.cherd.2010.01.022

186. V. Piemonte, L. Turchetti, M.C. Annesini, Bilirubin removal from albumin-containing solutions: dynamic adsorption on anionic resin. Asia-Pac. J. Chem. Eng. **5**(5), 708–713 (2010). doi:10.1002/apj.395

187. M.C. Annesini, V. Morabito, G. Novelli, V. Piemonte, L. Turchetti, Molecular adsorbent recirculating system (mars): a chemical engineering analysis of in vivo experimental data. Chem. Eng. Trans. **17**, 1095–1100 (2009b). doi:10.3303/CET0917183

188. M.C. Annesini, V. Piemonte, L. Turchetti, Artificial liver devices: a chemical engineering analysis. Asia-Pac. J. Chem. Eng. **6**(4), 639–648 (2011). doi:10.1002/apj.464

189. A.J. Wigg, R.T. Padbury, Liver support systems: promise and reality. J. Gastroenterol. Hepatol. **20**(12), 1807–1816 (2005). doi:10.1111/j.1440-1746.2005.03965.x

190. B. Struecker, N. Raschzok, I.M. Sauer, Liver support strategies: cutting-edge technologies. Nat. Rev. Gastroenterol. Hepatol. **11**(3), 166–176 (2014). doi:10.1038/nrgastro.2013.204

191. Yu. Cheng-Bo, X.-P. Pan, L.-J. Li, Progress in bioreactors of bioartificial livers. Hepatobiliary Pancreat. Dis. Int. **8**(2), 134–140 (2009)

192. M.E. Hoque, *Bioartificial liver assist system: tissue engineering challenges*. LAP Lambert Academic Publishing (2010)

193. G. Catapano, J.C. Gerlach, Bioreactors for liver tissue engineering. Top. Tissue Eng. **3**, 1–42 (2007)

194. G. Catapano, B.L. De, Combined effect of oxygen and ammonia on the kinetics of ammonia elimination and oxygen consumption of adherent rat liver cells. Int. J. Artif. Organs **25**(2), 151–157 (2002)

195. R. Anand Kumar, J.M. Modak, Transient analysis of mammalian cell growth in hollow fibre bioreactor. Chem. Eng. Sci. **52**(12), 1845–1860 (1997)

196. F. Coletti, S. Macchietto, N. Nlvassore, Mathematical modelling of three-dimensional cell cultures in perfusion bioreactors. part ii. Comput. Aided Chem. Eng. **21**, 1699–1704 (2006)

197. J. Park, F. Berthiaume, M. Toner, M.L. Yarmush, A.W. Tilles, Microfabricated grooved substrates as platforms for bioartificial liver reactors. Biotechnol. Bioeng. **90**(5), 632–644 (2005)

198. M. Iwamuro, H. Shiraha, S. Nakaji, K. Yamamoto, Prospects for creating bioartificial liver system with induced pluripotent stem cell technology. J. Biotechnol. Biomater. **3**(157), 2 (2013)

199. E. Adema, A.J. Sinskey, An analysis of intra-versus extracapillary growth in a hollow fiber reactor. Biotechnol. Prog. **3**(2), 74–79 (1987)

200. A.A. Demetriou, R.S. Brown, Jr., R.W. Busuttil, J. Fair, B.M. McGuire, P. Rosenthal, J.S.A. Esch, 2nd, J. Lerut, S.L. Nyberg, M. Salizzoni, E.A. Fagan, B. de Hemptinne, C.E. Broelsch, M. Muraca, J.M. Salmeron, J.M. Rabkin, H.J. Metselaar, D. Pratt, M. De La Mata, L.P. M.cChesney, G.T. Everson, P.T. Lavin, A.C. Stevens, Z. Pitkin, B.A. Solomon, Prospective, randomized, multicenter, controlled trial of a bioartificial liver in treating acute liver failure. Ann. Surg. **239**(5): 660–7; discussion 667–70 (2004)

201. S.L. Nyberg, R.A. Shatford, M.V. Peshwa, J.G. White, F.B. Cerra, W.S. Hu, Evaluation of a hepatocyte-entrapment hollow fiber bioreactor: a potential bioartificial liver. Biotechnol. Bioeng. **41**(2), 194–203 (1993). doi:10.1002/bit.260410205

202. J.C. Gerlach, J. Encke, O. Hole, C. Müller, J.M. Courtney, P. Neuhaus, Hepatocyte culture between three dimensionally arranged biomatrix-coated independent artificial capillary systems and sinusoidal endothelial cell co-culture compartments. Int. J. Artif. Organs **17**(5), 301–306 (1994)

203. E. Schmelzer, A. Finoli, I. Nettleship, J.C. Gerlach, Long-term three-dimensional perfusion culture of human adult bone marrow mononuclear cells in bioreactors. Biotechnol. Bioeng. 112(4), 801–810 (2015). doi:10.1002/bit.25485

204. L.M. Flendrig, R.A. Chamuleau, M.A. Maas, J. Daalhuisen, B. Hasset, C.G. Kilty, S. Doyle, N.C. Ladiges, G.G. Jörning, J.W. la Soe, D. Sommeijer, A.A. te Velde, Evaluation of a novel bioartificial liver in rats with complete liver ischemia: treatment efficacy and species-specific alpha-GST detection to monitor hepatocyte viability. J. Hepatol. 30(2), 311–320 (1999)

205. M.P. van de Kerkhove, E. Di Florio, V. Scuderi, A. Mancini, A. Belli, A. Bracco, M. Dauri, G. Tisone, G. Di Nicuolo, P. Amoroso, A. Spadari, G. Lombardi, R. Hoekstra, F. Calise, R.A.F.M. Chamuleau, Phase i clinical trial with the amc-bioartificial liver. Int. J. Artif. Organs 25(10), 950–959 (2002)

206. P.P.C. Poyck, G. Pless, R. Hoekstra, S. Roth, A.C.W.A. Van Wijk, R. Schwartländer, T.M. Van Gulik, I.M. Sauer, R.A.F.M. Chamuleau, In vitro comparison of two bioartificial liver support systems: mels cellmodule and amc-bal. Int. J. Artif. Organs 30(3), 183–191 (2007)

207. E. Doré, C. Legallais, A new concept of bioartificial liver based on a fluidized bed bioreactor. Ther. Apher. 3(3), 264–267 (1999)

208. Y.J. Hwang, Y.I. Kim, J.G. Lee, J.W. Lee, J.W. Kim, J.M. Chung, Development of bioartificial liver system using a fluidized-bed bioreactor. Transplant. Proc. 32(7), 2349–2351 (2000)

209. Y.J. Hwang, J.Y. Kim, S.K. Chang, S.G. Kim, Y.I. Kim, J.H. Hwang, Bioartificial liver system using a fluidized-bed bioreactor. Korean J. Hepato-Biliary-Pancreat. Surg. 12(1), 41–45 (2008)

210. B. David, E. Dore, M.Y. Jaffrin, C. Legallais, Mass transfers in a fluidized bed bioreactor using alginate beads for a future bioartificial liver. Int. J. Artif. Organs 27(4), 284–293 (2004)

211. B. Carpentier, A. Gautier, P. Paullier, C. Legallais, Mass transfer studies in a model of fluidized bed bioartificial liver, in XXXIII ESAO Congress (2006)

212. A. Kinasiewicz, A. Gautier, D. Lewinska, M. Dufresne, P. Paullier, J. Bukowski, C. Legallais, A. WeryNsky, Activity of cells within alginate beads produced for fluidized bed bioreactor (bioartificial liver), in XXIII ESAO Congress (2006)

213. C. Selden, C.W. Spearman, D. Kahn, M. Miller, A. Figaji, E. Erro, J. Bundy, I. Massie, S.-A. Chalmers, H. Arendse et al., Evaluation of encapsulated liver cell spheroids in a fluidised-bed bioartificial liver for treatment of ischaemic acute liver failure in pigs in a translational setting. PloS One 8(12), e82312 (2013)

214. A. Gautier, B. Carpentier, M. Dufresne, Q. Vu Dinh, P. Paullier, C. Legallais, Impact of alginate type and bead diameter on mass transfers and the metabolic activities of encapsulated c3a cells in bioartificial liver applications. Eur. Cell. Mater. 21, 94–106 (2011)

215. A. Di Martino, L. Liverani, A. Rainer, G. Salvatore, M. Trombetta, V. Denaro, Electrospun scaffolds for bone tissue engineering. Musculoskelet. Surg. 95(2), 69–80 (2011)